Civil Engineering

Sample Examination

Fifth Edition

Michael R. Lindeburg, PE

Professional Publications, Inc. • Belmont, CA

How to Get Online Updates for This Book

I wish I could claim that this book is 100% perfect, but 25 years of publishing have taught me that engineering textbooks seldom are. Even if you only took one engineering course in college, you are familiar with the issue of mistakes in textbooks.

I am inviting you to log on to Professional Publications' web site at **www.ppi2pass.com** to obtain a current listing of known errata in this book. From the web site home page, click on the "Errata" button. Every significant known update to this book will be listed as fast as we can say "HTML." Suggestions from readers (such as yourself) will be added as they are received, so check in regularly.

PPI and I have gone to great lengths to ensure that we have brought you a high-quality book. Now, we want to provide you with high-quality after-publication support. Please visit us at **www.ppi2pass.com**.

Michael R. Lindeburg, PE
Publisher, Professional Publications, Inc.

CIVIL ENGINEERING SAMPLE EXAMINATION
Fifth Edition

Printed in the United States of America

Professional Publications, Inc.
1250 Fifth Avenue, Belmont, CA 94002
(650) 593-9119
www.ppi2pass.com

Current printing of this edition: 1

ISBN 1-888577-60-6

The CIP data is pending.

Table of Contents

Preface to the Fifth Edition

Although the major subjects tested on the NCEES licensing exam for professional civil engineers remain essentially unchanged, the format of the exam has changed dramatically. This edition of the *Civil Engineering Sample Examination* reflects NCEES' change to a Breadth and Depth (B&D) format, implemented for civil engineers as of the October 2000 exam.

The B&D format actually ushers in several painful changes to the PE exam. The first change is that all questions on the exam are now multiple-choice. Gone are the free-response ("essay") problems where you could get partial credit based on your methodology. This can be a blessing in some problems, since all of the simplifying assumptions and required data must be provided. However, previous administrations of the exam have demonstrated pretty well that the multiple-choice questions are more difficult and time-consuming than the corresponding free-response questions.

The second change incorporated into the new format is the no-choice aspect of the morning session. This session consists of a broad collection of 40 questions, with no selection possible. To receive full credit for the morning, some expertise is required in all of the subject categories (structural and nonstructural). This appears to make the new exam more difficult than the abandoned format, which allowed examinees to pick four problems out of ten.

Finally, the new B&D format allows you to select from several afternoon specialty options: structural, transportation, water resources, environmental, and geotechnical. Although there might be some overlap of questions in these specialty options, presumably you will be able to work mostly in an area of expertise. The benefit of the privilege, however, is offset by the fact that the afternoon options have also been converted to no-choice, with no selection possible outside of the specialty options.

Engineering economics is a subject that continues to appear, here and there, once in a while. Although the subject has been eliminated from almost all of NCEES' PE exam outlines, NCEES allows that "Some problems will contain aspects of engineering economics." So, although gone are the days when an examinee had to exhibit the knowledge of a tax accountant, it is still open season for such questions.

NCEES has long asserted that it is very difficult to ask ten meaningful questions about one particular scenario. Apparently, there may have been a few "gimmees" in some of the ten-part multiple-choice questions of the past. The new format allows for scenarios with any number of multiple-choice questions.

NCEES has developed this new B&D format in order to reduce the fluctuations and eliminate the variations in the passing percentage, as well as to speed up, simplify, and economize on the grading and reporting processes. Another benefit to both NCEES and the state boards is in the reduction or altogether elimination of costly, time-consuming appeals.

None of the problems in this publication are actual exam problems. About 60% of the questions in this edition of the *Civil Engineering Sample Exam* are new questions that have never before been published. Most have come out of my head or are based on suggestions received from examinees over the years. The other 40% have been adapted from tried-and-true questions taken from the previous four editions of this publication.

In writing this book, I had some major decisions to make. Basically, I had to first decide if I wanted this sample exam to be 100% representative of the actual exam or if I wanted it to be useful to you as a review tool. I went 'round and 'round on this one. Eventually, I concluded that you wanted, more than anything else, to pass the PE exam, and that you could overlook minor variations from actual exam format as long as these variations (1) were pointed out to you, and (2) made this sample exam more effective. Most of the subsequent decisions followed directly from this first one.

For example, in this sample exam, each of the five afternoon specialty exams is based completely on the depth topic. In this sample exam, the geotechnical specialty exam contains 40 questions about soil. The actual PE exam will contain mostly questions about soil, but will mix in a little transportation, a little fluids, and so on. Same for the other specialty exams. I assumed that, if you were going to select a specialty, that you would want maximum exposure to that specialty. I decided, that if you wanted a random "smattering" of questions from some other specialty, you could just open to one of the other specialty exams. So, now, I've alerted you to the fact that you might have to exhibit a little knowledge of sludge bulking even though you have chosen to work the structures specialty exam.

The next decision involves the number of significant digits used in the solutions. The rule for significant digits is well-known: Answers cannot be any more precise than the most imprecise parameter. Field practice often uses even fewer significant digits than would be justified by the given data, in recognition of the many unknowns and assumptions incorporated into the solution.

But exam review is not field work. For this publication, I assumed you would be using a calculator and would want to compare this book's solutions with the digits appearing on your calculator. I also assumed that you would retain intermediate answers in your calculator's memory rather than entering new numbers that you have already manually (or mentally) rounded. Because of these assumptions, most of the results are printed with more significant digits than can be justified.

Another decision involves the units used in the solutions. I used common field units wherever possible, but not everywhere. For example, I tried to be consistent with ACI 318 and the AISC *Manual of Steel Construction*. Therefore, I may have used "kips" in one place, "k" in another, and "pounds" in yet another. Similarly, one problem might have been solved using "ksi" while another was solved with "psi." It depended on the source document.

I have used the same variable symbols as in the *Civil Engineering Reference Manual*, which was modeled after the original source documents as well.

In keeping with a personal preference, I was strict in differentiating between weight and mass. Thus, the units "lbf" and "lbm" are differentiated throughout, although "lb" is good enough for most civil engineering work. Similarly, I differentiated mass density (ρ) from specific weight (γ), even though the good-old "pcf" is comfortably indistinct in its meaning.

When writing and solving the few metric problems in this book, I used only standard SI units. You won't find any "kilogram-force" units in this book, even though they might still be used in other countries. The SI system uses the kilogram as a unit of mass, never as weight or force.

NCEES makes an attempt to "decouple" subsequent questions from previous results. For example, if question 1 asks, "What is the coefficient of active earth pressure?", then question 2 will often be phrased, "Assuming the coefficient of active earth pressure is 0.30, what is the active force on the wall?" In this manner, you still have a chance on question 2 even if you get 1 wrong. That's good.

And it's bad. In some problem sets, this decoupling results in your having to repeat previous steps with the new set of given data for each subsequent question answered. Some examinees have complained about the additional workload and not being able to use the interim results in their calculator stack; many others are quite comfortable with this. I used decoupling sparingly in this edition, to be respectful of your time before the exam.

Finally, this sample exam is a little easier—about 10%—than the actual exam. It's still representative and realistic, just a tiny bit easier. I admit it. My sample exams always have been, because I plan it that way. Think about it: You're taking the sample exam one or two weeks before the actual exam. You studied as much as you could. Then, Mr. Lindeburg throws a collection of the most obscure problems he could find at you? Stuff that you didn't study? This is going to accomplish what? Anything? Email me if you have an answer and want me to toughen up the problems.

Everyone is different; everyone has different strengths and weaknesses. Everyone has a different knowledge base and working speed. Not to be forgotten is that the passing rate for the PE exam has never been consistent, sometimes doubling or halving from one administration to the next. With each examinee being different, and each PE exam being different, it is unlikely that I would ever be able to perfectly match the complexity and level of difficulty you experience anyway.

I don't think one engineer in a hundred is going to be concerned with my editorial decisions. However, I made them, and I wanted you to know about them so that you aren't surprised when you see the real exam is different. This sample exam is a dry-run tool, not a licensing exam.

Inasmuch as I, apparently, am very human, there are going to be errata associated with this book—particularly in the first few printings. Errata will be published online as soon as they have been discovered and verified. PPI has established an errata web page, accessible from PPI's home page at www.ppi2pass.com, to keep the readers updated.

As in all of my publications, I welcome your comments. If you disagree with a solution or if you think there is a better way to do something, please let me know so that I can share your comments with everyone else. You can email me directly, or use one of the business reply cards at the back of this book.

Michael R. Lindeburg, PE
mlindeburg@ppi2pass.com

Acknowledgments

I think my first final exam consisted of about eight pages. In the good old days, I could sit down and type up the final exam for one of my classes on ditto masters. The solutions would be hand-scribbled on other ditto masters. Nobody expected anything else.

But technology didn't stand still. Following the strike-on ditto master came heat- and photo-sensitive ditto masters. This amazing technology enabled me to prepare a master on regular white paper, covering up my errors either with a toxic error-correction fluid or by retyping the error through one of the pieces of strike-on correction paper that seemed to permanently "live" in a plastic tray attached to the side of every IBM Selectric typewriter. There was a similar spark-assisted mimeograph process that let you prepare masters for longer runs than could be accommodated by ditto sheets.

Then came copy machines that used a tissue paper much like that used by the original fax machines. With their special document "carriers," these machines were too slow and cumbersome to be used for classroom publishing, let alone national distribution. And, besides, they had an uncanny way of blurring even the most precise marks on the original. The "advanced" wet-process copy machines of the day not only required you to hang your copies to dry before using them, but they used a coated paper that seemed to deny attempts to mark on them with any writing implement. Rather than permit a visible mark, such wet-process papers would sacrifice themselves by creating a gooey mess on the end of your pencil. And, Heaven forbid should you ever need to use white-out on some of that coated paper. Even subsequent dry photocopiers (i.e., "Xerox machines," sic) did not change the method by which the original document was prepared.

Over the years, I've "matured" from hand-scribbling to a wedding present of a Panasonic typewriter with two (yes, two!) interchangeable keys, to an IBM Selectric with dozens of font balls, to electronic word processors (a term that now refers to a person rather than to a piece of equipment) with replaceable font wheels, to various specialty typewriters with miniature fonts, to cold-copy typesetting equipment such as the Varityper and IBM Composer—some with card memory, some with cassette memory, and some with human memory. All of these methods were awkward, to say the least, when it came to typesetting mathematical and scientific symbols. (I think I may still have the world's largest supply of rub-on symbols!)

It wasn't until Donald Knuth developed $\TeX$ and gave it to the world that it became possible to typeset really beautiful engineering and mathematical books. When $\TeX$ first became available, it was only on the Stanford Computer System, which we accessed from about 10 miles away from campus with a 300-, then 1200-, and finally (gasp!) a 2400-baud modem that cost $1000. We've been using $\TeX$ ever since.

Illustrations have evolved similarly, though that's a story I'm leaving for another book. Suffice to say, you can't even find people today who can pronounce "Rapidograph pen," let alone take one apart to unclog it.

Anyway, that's the way things were published in the past. Nowadays, Professional Publications has a production department that other publishers would kill for. The competence, dedication, professionalism, and methodologies of this group are the antitheses of the early days I described.

So, thank you, denizens of the production department: Jessica Holden (copyediting); Kate Hayes (typesetting); Yvonne Sartain (illustrating and cover design); Tracey Brown (proofreading); and Cathy Schrott (coordination and administration). Without you, PPI customers would be holding sheafs of stapled, wet, smelly pages with funny blue letters.

From a technical standpoint, thanks also to PPI's staff engineer, Tamer Hadi, for doing the in-house technical check of this manuscript.

What a team!

Michael R. Lindeburg, PE

How to Use this Book

This book is a sample exam, so there are only a few ways that you can use it. Some people will work through every question, basically using this book as a collection of solved problems. At the other extreme, some people will run out of time and won't use it at all. But, to me, the main issue is not *how* or when you use this sample exam, but *what* you learn from it.

Though I tried to include realistic exam problems, I did not write this book intending it to be a diagnostic tool guiding your preparation. You shouldn't take this sample exam and then design your review around what you didn't know. If you take this exam and don't do well on a particular problem, I wouldn't want you to spend the next three months preparing for that type of problem. The tried-and-true method of exam preparation is a systematic, thorough, and complete approach based on long-term exam trends, not based on transient and oddball fads. The tried-and-true study method is what the *Civil Engineering Reference Manual* is for.

The problems in this sample exam were written for several reasons, but providing you with problems and solutions you could copy on the exam was not one of them. You should be aware that it is disappointingly easy for actual exam questions to either (a) subtlely diverge from the assumptions made in a solution in this sample exam, or (b) negate the entire point of what you learned from a problem in this sample exam. If you should find an exam problem similar to something in one of my publications, don't count yourself lucky.

Similarly, you mustn't read too much into what you don't see or don't need to know in this sample exam. For example, there may be no problems that require you to access the AASHTO "Green Book." Does that mean you shouldn't take the Green Book to the exam? Definitely not!

The value of a sample exam does not lie in its ability to guide your preparation. Rather, the value is in giving you an opportunity to bring together all of your knowledge and to practice your test-taking skills. The three most important skills are (1) familiarizing yourself with the exam structure, (2) organizing your references and other resources, and (3) managing your time. I intended this sample exam to be taken within a few weeks of your actual exam. That's the only time that you will be able to focus on test-taking skills without the distraction of rusty recall.

You'll need to set aside an entire day in order to take this sample exam the way I intended it to be used. I know using up another day is asking a lot from you. But if you start early enough and study diligently, by the time the actual exam rolls around, you will probably be weak in only one area: familiarity with the nature of the exam.

I know that you are going to be frustrated when you discover that you need more than the *Civil Engineering Reference Manual* to score 100% on this sample exam. And you are going to find that some questions will "cost" you about 30 seconds of time, whereas others will cost you 5 to 10 minutes. However, the average time will be adequate. Also, you will find the spread of answer choices annoyingly wide. Your answer—the answer that you "know" is correct—just isn't among the available choices. You'll just have to select the closest answer. All of these characteristics are representative of the actual exam. This sample exam will do little to ensure confidence in your preparation. The actual exam will similarly keep you guessing about your outcome up to the day you receive your results.

The other things that I hope you learn from this sample exam are: (1) questions are sometimes worded poorly, distractingly, and strangely; (2) many scenarios initially have misleadingly simple questions that lead you into more difficult interiors; and (3) most of the problems have some tricky parts that can derail you even if you know the subject material. The exam isn't diabolical; it isn't trying to get you to fail. However, the exam is intended to separate those who "know their stuff" from others who only have a superficial knowledge of the fluff.

One always-vexing difficulty concerns which edition or version of the codes to use. The latest codes, as listed in the following section, were used for this sample exam. The actual exam may not be as current as this sample exam. (Those of you who try to use this sample exam long after its useful life may not find anything that you recognize.)

PPI has published the current code versions and editions on its web site for the civil exam. You should select your references on the basis of what will be tested on the actual exam, not on what this sample exam has used. If the codes are different, all you need to know is how to solve a particular type of problem using the code being tested for.

In athletics, coaches often speak of the home-field advantage. Athletes who are comfortable in their environment play better. Well, examinees who have "seen it before" via a sample exam have a psychological edge as well. This publication was written to give you that edge.

Good luck!

Codes Used to Prepare this Book

This edition of the *Civil Engineering Sample Exam* is based on the codes, standards, regulations, and references listed here. The versions and editions (i.e, dates) that I used were the most current available to me. However, as with engineering practice itself, the NCEES PE examination is not always based on the most current codes, as adoptions by state and local agencies often lag issuance by several years.

On its web site, PPI lists the dates of the codes, standards, regulations, and references upon which NCEES bases the current exam. You should use this information to decide which versions of these books should be part of your exam preparation.

Actual building codes may be needed, but only to a limited extent (e.g., for wind and seismic loadings). Since the PE exam is a national exam, NCEES cannot require you to use one particular building code. So, UBC, SBCCI, and BOCA versions of all code-related questions are presented. The code edition (year, version, etc.) tested is a thorny issue, however. The exam can be years behind the most recently adopted code, and it may be many, many years behind the latest codes being published and distributed.

Policy on Geometric Design of Highways and Streets ("AASHTO Green Book") English units version, 1990. American Association of State Highway and Transportation Officials (AASHTO).

Highway Capacity Manual, Special Report 209, 1997. Transportation Research Board/National Research Council.

Metric Analysis Guide, Supplement to 1997 Update of Special Report 209 of the Highway Capacity Manual, 1997. Transportation Research Board/National Research Council.

Building Code Requirements for Structural Concrete (ACI-318) with Commentary (ACI-318R), 1995. American Concrete Institute, Farmington Hills, MI.

Notes on ACI 318-95: Building Code Requirements for Structural Concrete with Design Applications, 1996. Portland Cement Association.

Design Handbook in Accordance with the Strength Design Method of ACI-318 (ACI SP-17), 1990. American Concrete Institute, Detroit, MI.

Standard Specifications for Highway Bridges, 1992, with 1993 and 1994 interim specification updates. American Association of State Highway and Transportation Officials (AASHTO).

Manual of Steel Construction, Allowable Stress Design, 1989. American Institute of Steel Construction, Inc., 9th ed.

Load and Resistance Factor Design, Vol. 1, "Structural Members, Specifications, and Codes," 2nd ed., 1993. American Institute of Steel Construction, Inc. (AISC).

Uniform Building Code, Structural Engineering Design Provisions, Vol. 2, 1997. International Conference of Building Officials (ICBO).

Manual of Uniform Traffic Control Devices—For Streets and Highways (MUTCD), D6.1, 1988. U.S. Department of Transportation, Federal Highway Administration (FHA), Traffic Control Systems Divisions, Office of Traffic Operations.

Roadside Design Guide, 1996. American Association of State Highway and Transportation Officials (AASHTO).

Thickness Design for Concrete Highway and Street Pavements, 1984 (reprinted in 1995). Portland Cement Association, Skokie, IL.

AASHTO Guide for Design of Pavement Structures, Vol. 1, 1993. American Association of State Highway and Transportation Officials (AASHTO).

AASHTO Guide for Design of Pavement Structures, Vol. 2, 1986. American Association of State Highway and Transportation Officials (AASHTO).

The Asphalt Handbook, Manual MS-4, 1988. The Asphalt Institute.

Recommended Standard for Wastewater Facilities ("10 States' Standards," TSS), 1997. Health Education Services, Health Resources, Inc., Albany, NY.

Standard Methods for the Examination of Water and Wastewater, 19th ed., 1995, a joint publication of the American Public Health Association (APHA), the American Water Works Association (AWWA), and the Water Pollution Control Federation (WPCF).

About the Civil PE Exam

The civil PE exam is eight hours long, with two four-hour sessions, morning and afternoon. All problems are multiple-choice. This exam is divided into a "breadth and depth" format. In the morning session, all examinees work the same "breadth" exam, which consists of 40 problems drawn from all five areas of civil engineering listed below. In the afternoon, examinees choose to work one of five "depth" exams: Environmental, Geotechnical, Structural, Transportation, or Water Resources. Each depth exam consists of 40 problems that test knowledge in the areas specified.

MORNING SESSION
(40 multiple-choice problems)

(1) Environmental—approx. 20% of problems

- Wastewater Treatment: wastewater flow rates, unit processes
- Biology: toxicity, algae, stream degradation, temperature, disinfection, water taste and odor, BOD
- Solid/Hazardous Waste: collection, storage/transfer, treatment, disposal, quantity estimates, site
- Groundwater and Well Fields: groundwater flow, aquifers (e.g., characterization)

(2) Geotechnical—approx. 20% of problems

- Subsurface Exploration and Sampling: drilling and sampling, soil classification, boring log interpretation, soil-profile development
- Engineering Properties of Soils: index properties, phase relationships, permeability
- Soil Mechanics Analysis: pressure distribution, lateral earth pressure, consolidation, compaction
- Shallow Foundations: bearing capacity, settlement, allowable bearing pressure
- Earth Retaining Structures: gravity walls, cantilever walls, earth pressure diagrams, stability analysis

(3) Structural—approx. 20% of problems

- Loadings: dead and live loads, wind loads
- Analysis: determinate analysis, shear diagrams, moment diagrams

- Mechanics of Materials: flexure, shear, tension and compression, deflection
- Materials: reinforced concrete, structural steel, timber, concrete mix design, masonry
- Member Design: beams, slabs, columns, reinforced concrete footings, retaining walls, trusses

(4) Transportation—approx. 20% of problems

- Traffic Analysis: capacity analysis
- Construction: excavation/embankment, material handling, optimization, scheduling
- Geometric Design: horizontal curves, vertical curves, sight distance

(5) Water Resources—approx. 20% of problems

- Hydraulics: energy dissipation, energy/continuity equation, pressure conduit, open-channel flow, flow rates, friction/minor losses, flow equations, hydraulic jump, culvert design, velocity control
- Hydrology: storm characterization, storm frequency, hydrographs, rainfall intensity and duration, runoff analysis
- Water Treatment: demands, hydraulic loading, storage (raw and treated water)

Note: NCEES states that these areas are examples of the kinds of knowledge that will be tested but are not exclusive or exhaustive categories.

AFTERNOON SESSION

Examinees must choose to work one of the following five depth exams. Each depth exam has 40 multiple-choice problems, and examinees must work all problems.

Civil/Environmental Depth Exam
(40 multiple-choice problems)

(1) Environmental—approx. 65% of problems

- Wastewater Treatment: wastewater flow rates, primary clarification, biological treatment, secondary clarification, chemical precipitation, sludge systems, digesters, disinfection, nitrification/denitrification, effluent limits, wetlands, unit processes, operations

- Biology (including micro and aquatic): toxicity, algae, food chain, stream degradation, organic load, oxygenation / deoxygenation / oxygen sag curve, eutrophication, temperature, indicator organisms, disinfection, water taste and odor, most probable number (MPN), BOD, quality control

- Solid/Hazardous Waste: collection, storage/transfer, treatment, disposal, quantity estimates, site and haul economics, energy recovery, hazardous waste systems, applicable standards

- Groundwater and Well Fields: dewatering, well analysis, water quality analysis, subdrain systems, groundwater flow, groundwater contamination, recharge, aquifers (e.g., characterization)

(2) Geotechnical—approx. 10% of problems

- Subsurface Exploration and Sampling: drilling and sampling procedures, soil classification, boring log interpretation, soil-profile development

- Engineering Properties of Soils: permeability

- Soil Mechanics Analysis: compaction, seepage and erosion

(3) Water Resources—approx. 25% of problems

- Hydraulics: energy/continuity equation, pressure conduit, open channel flow, detention/retention ponds, pump application and analysis, pipe network analysis, flow rates (domestic, irrigation, fire), surface water profile, cavitation, friction/minor losses, flow measurement devices, flow equations, culvert design, velocity control

- Hydrology: storm characterization, storm frequency, hydrograph (unit and others), transpiration, evaporation, permeation, rainfall intensity and duration, runoff analysis, gauging stations, flood plain/floodway, sedimentation

- Water Treatment: demands, hydraulic loading, storages (raw and treated water), rapid mixing, flocculation, sedimentation, filtration, disinfection, applicable standards

Note: NCEES states that these areas are examples of the kinds of knowledge that will be tested but are not exclusive or exhaustive categories.

Civil/Geotechnical Depth Exam
(40 multiple-choice problems)

(1) Geotechnical—approx. 65% of problems

- Subsurface Exploration and Sampling: drilling and sampling procedures, in-situ testing, soil classification, boring log interpretation, soil-profile development

- Engineering Properties of Soils: index properties, phase relationships, shear strength properties, permeability

- Soil Mechanics Analysis: effective and total stresses, pore pressure, pressure distribution, lateral earth pressure, consolidation, compaction, slope stability, seepage and erosion

- Shallow Foundations: bearing capacity, settlement, allowable bearing pressure, proportioning individual/combined footings, mat and raft foundations, pavement design

- Deep Foundations: axial capacity (single pile/drilled shaft), lateral capacity (single pile/drilled shaft), settlement, lateral deflection, behavior of pile/drilled shaft groups, pile dynamics and pile load tests

- Earth Retaining Structures: gravity walls, cantilever walls, mechanically stabilized earth wall, braced and anchored excavations, earth dams, earth pressure diagrams, stability analysis, serviceability requirements

- Seismic Engineering: earthquake fundamentals, liquefaction potential evaluation

(2) Environmental—approx. 10% of problems

- Groundwater and Well Fields: dewatering, water quality analysis, groundwater contamination, aquifers (e.g. characterization)

(3) Structural—approx. 20% of problems

- Loadings: dead and live loads, earthquake loads

- Materials: concrete mix design

- Member Design: reinforced concrete footings, pile foundations, retaining walls

(4) Transportation—approx. 5% of problems

- Construction: excavation/embankment, pavement design

Note: NCEES states that these areas are examples of the kinds of knowledge that will be tested but are not exclusive or exhaustive categories.

Civil/Structural Depth Exam
(40 multiple-choice problems)

(1) Structural—approx. 65% of problems

- Loadings: dead and live loads, moving loads, wind loads, earthquake loads, repeated loads

- Analysis: determinate, indeterminate, shear diagrams, moment diagrams

- Mechanics of Materials: flexure, shear, torsion, tension and compression, combined stresses, deflection

- Materials: reinforced concrete, pre-stressed concrete, structural steel, timber, concrete mix design, masonry, composite construction
- Member Design: beams, slabs, columns, reinforced concrete footings, pile foundations, retaining walls, trusses, braces and connections, shear and bearing walls
- Failure Analysis: buckling, fatigue, failue modes
- Design Criteria: UBC, BOCA, SBC, ACI, PCI, AISC, AITC, AASHTO, ASCE-7

(2) Geotechnical—approx. 25% of problems

- Subsurface Exploration and Sampling: boring log interpretation
- Soil Mechanics Analysis: pressure distribution, lateral earth pressure
- Shallow Foundations: bearing capacity, settlement, proportioning individual/combined footings, mat and raft foundations
- Deep Foundations: axial capacity (single pile/drilled shaft), lateral capacity (single pile/drilled shaft), behavior of pile/drilled shaft groups
- Earth Retaining Structures: gravity walls, cantilever walls, braced and anchored excavations, earth pressure diagrams, stability analysis

(3) Transportation—approx. 10% of problems

- Construction: excavation/embankment, material handling, optimization, scheduling

Note: NCEES states that these areas are examples of the kinds of knowledge that will be tested but are not exclusive or exhaustive categories.

Civil/Transportation Depth Exam
(40 multiple-choice problems)

(1) Transportation—approx. 65% of problems

- Traffic Analysis: traffic signal, speed studies, capacity analysis, intersection analysis, parking operations, traffic volume studies, mass transit studies, sight distance, traffic control devices, pedestrian facilities, bicycle facilities, driver behavior/performance
- Transportation Planning: origin-destination studies, site impact analysis, capacity analysis, optimization/cost analysis, trip generation/distribution/assignment
- Construction: excavation/embankment, material handling, optimization, scheduling, mass diagrams, pavement design
- Geometric Design: horizontal curves, vertical curves, sight distance, superelevation, vertical/horizontal clearances, acceleration and deceleration, intersections/interchanges

- Traffic Safety: accident analysis, roadside clearance analysis, counter-measurement development, economic analysis, conflict analysis

(2) Geotechnical—approx. 15% of problems

- Subsurface Exploration and Sampling: soil classification, boring log interpretation, soil-profile development
- Engineering Properties of Soils: index properties, phase relationships
- Soil Mechanics Analysis: compaction, seepage and erosion
- Shallow Foundations: pavement design

(3) Water Resources—approx. 20% of problems

- Hydraulics: open-channel flow, flow rates (domestic, irrigation, fire), flow equations, culvert design, velocity control
- Hydrology: rainfall intensity and duration, runoff analysis, flood plain/floodway

Note: NCEES states that these areas are examples of the kinds of knowledge that will be tested but are not exclusive or exhaustive categories.

Civil/Water Depth Exam
(40 multiple-choice problems)

(1) Water Resources—approx. 65% of problems

- Hydraulics: spillway capacity, energy dissipation, energy / continuity equation, pressure conduit, open-channel flow, detention / retention ponds, pump application and analysis, pipe network analysis, stormwater collection, flow rates (domestic, irrigation, fire), surface water profile, cavitation, friction/minor losses, sub- and supercritical flow, hydraulic jump, flow measurement devices, flow equations, culvert design, velocity control
- Hydrology: storm characterization, storm frequency, hydrographs (unit and others), transpiration, evaporation, permeation, rainfall intensity and duration, runoff analysis, gauging stations, flood plain/floodway, sedimentation
- Water Treatment: demands, hydraulic loading, storage (raw and treated water), rapid mixing, flocculation, sedimentation, filtration, disinfection, applicable standards

(2) Environmental—approx. 25% of problems

- Wastewater Treatment: unit processes
- Biology (including micro and aquatic): toxicity, algae, food chain, stream degradation, organic load, eutrophication, temperature, indicator organisms, disinfection, water taste and odor, most probable number (MPN), BOD, quality control

- Groundwater and Well Fields: well analysis, water quality analysis, groundwater flow, groundwater contamination, recharge, aquifers (e.g., characterization)

(3) Geotechnical—approx. 10% of problems

- Subsurface Exploration and Sampling: soil classification, boring log interpretation, soil-profile development
- Engineering Properties of Soils: permeability
- Soil Mechanics Analysis: seepage and erosion

Note: NCEES states that these areas are examples of the kinds of knowledge that will be tested but are not exclusive or exhaustive categories.

Instructions

In accordance with the rules established by your state, you may use textbooks, handbooks, bound reference materials, and any approved battery- or solar-powered, silent calculator to work this examination. However, no blank papers, writing tablets, unbound scratch paper, or loose notes are permitted. Sufficient room for scratch work is provided in the Examination Booklet.

You are not permitted to share or exchange materials with other examinees. However, the books and other resources used in this morning session may be changed prior to the afternoon session.

You will have four hours in which to work this session of the examination. Your score will be determined by the number of questions that you answer correctly. There is a total of 40 questions, divided into sets of two to six questions, each set pertaining to a particular scenario. All 40 questions must be worked correctly in order to receive full credit on the exam. There are no optional questions. Each question is worth 1 point. The maximum possible score for this section of the examination is 40 points.

Partial credit is not available. No credit will be given for methodology, assumptions, or work written in your Examination Booklet.

Record all of your answers on the Answer Sheet. No credit will be given for answers marked in the Examination Booklet. Mark your answers with a no. 2 pencil. Answers marked in pen may not be graded correctly. Marks must be dark and must completely fill the bubbles. Record only one answer per question. If you mark more than one answer, you will not receive credit for the question. If you change an answer, be sure the old bubble is erased completely, as incomplete erasures may be misinterpreted as answers.

If you finish early, check your work and make sure that you have followed all instructions. After checking your answers, you may turn in your Examination Booklet and Answer Sheet and leave the examination room. Once you leave, you will not be permitted to return to work or change your answers.

When permission has been given by your proctor, break the seal on the Examination Booklet. Check that all pages are present and legible. If any part of your Examination Booklet is missing, your proctor will issue you a new Booklet.

Do not work any questions from the Afternoon Session during the first four hours of this exam.

WAIT FOR PERMISSION TO BEGIN

Name: _____
      Last   First   Middle Initial

Examinee number: _____

Examination Booklet number:_____

Principles and Practice of Engineering Examination

Morning Session
Sample Examination

Morning Session

11:41

SITUATION FOR PROBLEMS 1-5

The NRCS (SCS) Curve Number method is being used to evaluate the runoff characteristics of a residential development constructed in a small basin in Arizona. The basin is divided into three subbasins. Subbasin 1 consists of townhouses, each with $\frac{1}{8}$ of an acre or less. Subbasin 2 consists of larger ranch-style homes, each with $\frac{1}{2}$ acre. Subbasin 3 is undeveloped natural desert. The subbasin areas, points of concentration, critical elevations, lengths of longest flowpaths, and coefficients of peak discharge are shown in the following illustration.

HM 0196

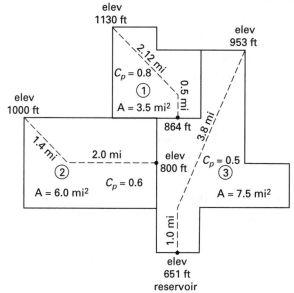

1. For the purpose of the NRCS (SCS) methodology, what is the average watershed slope for subbasin 1?

  (A) 0.19%

  (B) 0.95%

  (C) 1.9%

  (D) 2.4%

2. What NRCS (SCS) hydrologic soil group (HSG) best describes basin 1?

  (A) A

  (B) B

  (C) C

  (D) D

3. Given antecedent runoff condition II, what NRCS (SCS) curve number, CN, should be used for subbasin 1?

  (A) 63

  (B) 77

  (C) 85

  (D) 90

4. According to the NRCS (SCS) method, what is the soil storage capacity (i.e., the potential maximum retention after runoff begins) for subbasin 1?

  (A) 1 in

  (B) 2 in

  (C) 3 in

  (D) 4 in or more

5. What is the NRCS (SCS) time lag for subbasin 1?

  (A) 2 min

  (B) 15 min

  (C) 30 min

  (D) 120 min

SITUATION FOR PROBLEMS 6-10

A36 W 8 × 48 steel girders are placed on 5 ft centers to support a 4 in thick concrete floor slab. Shear connectors are used so that the girders/slab systems act together to resist bending. Each girder is 25 ft long and can be considered to be simply supported. The concrete compressive strength is 4000 psi. The girders are temporarily shored during construction.

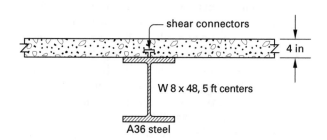

6. The effective flange width of an interior beam is closest to

(A) 3 ft

(B) 4 ft

(C) 5 ft

(D) 6 ft

7. Using the traditional Whitney assumptions, the ultimate normal force in the steel is most nearly

(A) 250,000 lbf

(B) 390,000 lbf

(C) 440,000 lbf

(D) 510,000 lbf

8. Using the traditional Whitney assumptions, the height of the compressive stress block in the concrete slab is most nearly

(A) 2.5 in

(B) 3.0 in

(C) 3.5 in

(D) 4.0 in

9. The distance between the tensile resultant, F_T, and the compressive resultant, F_C, is most nearly

(A) 6.5 in

(B) 7.0 in

(C) 8.0 in

(D) 8.5 in

10. Using ACI 318, what approximate uniform live load per foot of girder support must be placed on the slab in order for an interior slab to reach its ultimate moment?

(A) 1200 lbf/ft

(B) 1500 lbf/ft

(C) 1800 lbf/ft

(D) 2100 lbf/ft

SITUATION FOR PROBLEMS 11–15

A square, reinforced concrete footing is installed so that the footing bearing surface is 5 ft below the soil level, at a point where the allowable soil pressure is 3500 psf. Other than the soil above the footing, there is no surcharge. The soil unit weight is 100 lbf/ft³. The footing is located at the corner of a building and is loaded through a concentric 14 in square column. The column transmits a 125,000 lbf service dead load and a 175,000 lbf service live load to the footing. The dead

load includes the column's weight but does not include the footing's weight. The compressive strength for all of the concrete used is 3000 psi.

11. What is the minimum footing size?

(A) 9 ft × 9 ft

(B) 10 ft × 10 ft

(C) 11 ft × 11 ft

(D) 12 ft × 12 ft

12. The critical (plan) area contributing to two-way punching shear is closest to

(A) 60 ft² or less

(B) 70 ft²

(C) 80 ft²

(D) 90 ft² or more

13. The nominal concrete shear strength resisting punching shear is most nearly

(A) 350,000 lbf

(B) 450,000 lbf

(C) 550,000 lbf

(D) 650,000 lbf

14. What is the ultimate two-way punching shear?

(A) 350,000 lbf

(B) 450,000 lbf

(C) 550,000 lbf

(D) 650,000 lbf

15. What is the critical area for beam-action shear forces?

(A) 20 ft²

(B) 30 ft²

(C) 40 ft²

(D) 50 ft²

SITUATION FOR PROBLEMS 16–18

A, B, C, and D represent locations along various straight highway sections. AB represents the number of daily trips from location A to location B, while BA represents the number of daily trips from B to A. Other combinations of A, B, C, and D are interpreted similarly. All proposed interchanges connect rural interstate highways, and there are no other interchanges closer than 3.7 mi.

16. Given the following ADT traffic movements, what type of interchange would you recommend?

AB 18,500	AC 25	AD 15
BA 17,000	BD 30	BC 10
CD 90	CA 15	CD 20
DC 120	DB 25	DA 20

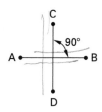

(A)   cloverleaf, full

(B)   cloverleaf, partial

(C)   diamond

(D)   directional

17. Given the following ADT traffic movements, what type of interchange would you recommend?

AB 15,000	AC 2400	AD 2600
BA 16,500	BD 1500	BC 1800
CD 12,000	CA 3600	CD 2200
DC 14,500	DB 1900	DA 3200

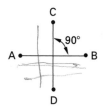

(A)   cloverleaf, full

(B)   cloverleaf, partial

(C)   diamond

(D)   directional

18. Given the following ADT traffic movements, what type of interchange would you recommend?

AB 25,000	AC 12,500	AD 400
BA 19,000	BD 9000	BC 3800
CD 15,500	CA 19,000	CD 5000
DC 10,000	DB 7500	DA 650

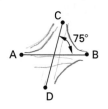

(A)   cloverleaf, full

(B)   cloverleaf, partial

(C)   diamond

(D)   directional

SITUATION FOR PROBLEMS 19 AND 20

A culvert system is being designed to pass under a major highway. The culvert system must be able to protect the highway from runoff from a 2.5 hr design storm. The following information has been derived from a storm that produced runoff over a 2.0 hr period.

drainage area	43 mi^2
flood hydrograph peak discharge	9300 ft^3/sec
flood hydrograph volume	3260 ac-ft

19. What is the 2 hr unit hydrograph peak discharge?

(A)   2300 ft^3/sec

(B)   3300 ft^3/sec

(C)   4800 ft^3/sec

(D)   6500 ft^3/sec

20. What is the design flood hydrograph volume?

(A)   4900 ac-ft

(B)   5700 ac-ft

(C)   7100 ac-ft

(D)   9800 ac-ft

SITUATION FOR PROBLEMS 21–25

A water treatment plant has the option of adding various compounds to its raw supply water. The effects on subsequent processes of such compounds are being evaluated. The compounds being considered are

  I.   NaOH
 II. CO_2
III. Na_2CO_3
 IV.   HCl

KA 0997

21. Which of the compounds, if added by itself, will increase the total inorganic carbon content?

(A)   II

(B)   III

(C)   II and III

(D)   II, III, and IV

22. Which of the compounds, if added by itself, will increase the pH of the raw water?

(A) I

(B) II and IV

(C) I, II, and III

(D) IV

23. Which of the compounds, if added by itself to the supply, will affect the alkalinity of the water?

(A) I and IV

(B) II and III

(C) I, III, and IV

(D) I, II, III, and IV

24. What will be the effect of adding only Na_2CO_3 to the supply water?

(A) decrease alkalinity, no effect on pH, increase inorganic carbon

(B) decrease alkalinity, decrease pH, decrease inorganic carbon

(C) increase alkalinity, decrease pH, increase inorganic carbon

(D) increase alkalinity, no effect on pH, increase inorganic carbon

25. What will be the effect of adding only HCl to the supply water?

(A) decrease alkalinity, increase pH, decrease inorganic carbon

(B) decrease alkalinity, decrease pH, no effect on inorganic carbon

(C) increase alkalinity, increase pH, no effect on inorganic carbon

(D) increase alkalinity, decrease pH, increase inorganic carbon

SITUATION FOR PROBLEMS 26–30

A 10 MGD wastewater treatment plant is being designed to process influent with a raw suspended solids content of 500 mg/L. The target suspended-solids concentration leaving the primary clarifier treatment is 150 mg/L. A pilot plant has been built with the results shown.   *KA 0797*

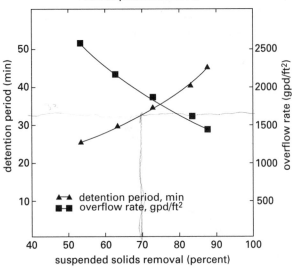

detention period and overflow rate versus percent removal

26. What is the approximate required efficiency of the primary clarification process?

(A) 30%

(B) 45%

(C) 70%

(D) 85%

27. What is the design flow rate?

(A) 300,000 gal/hr

(B) 400,000 gal/hr

(C) 650,000 gal/hr

(D) 1,200,000 gal/hr

28. Assume one circular clarifier with a capacity of 15 MGD is to be constructed. Assume a suspended solids removal efficiency of 70%. What is the required diameter?

(A) 40 ft

(B) 55 ft

(C) 70 ft

(D) 100 ft

29. Assume two units with 72 ft diameters will be built with a combined capacity of 15 MGD. Assume a suspended solids removal efficiency of 70%. What is the tank depth?

(A) 4 ft

(B) 6 ft

(C) 8 ft

(D) 10 ft

30. Assume two units with 72 ft diameters will be built having a combined capacity of 15 MGD. The weir loading rate is approximately

(A) 30,000 gal/day-ft

(B) 40,000 gal/day-ft

(C) 50,000 gal/day-ft

(D) 60,000 gal/day-ft

SITUATION FOR PROBLEMS 31–35

A geotechnical evaluation has been carried out at the corner of Jessica Ave. and Emily St. A page from the completed report log is shown.

31. How many total holes were drilled during the evaluation of the site?

(A) 1 or more

(B) 2 or more

(C) 3 or more

(D) 4 or more

32. The designation "SS" in the "No. & Type" column most likely means

(A) split spoon

(B) saturated sample

(C) saturated surface

(D) submerged sample

TEST HOLE LOG

Hole No. _____3_____   Site JESSICA AVE. & EMILY ST.

Date drilled __02-08-07__   Elevation _____604_____

33. The soil found at a depth of 1.0 m was classified as

(A) silty clay
(B) silty sand
(C) sandy silt
(D) sandy clay till

34. What description would be given to the soil at a depth of 2.5 m?

(A) loose
(B) medium dense
(C) dense
(D) very dense

35. What is the undrained strength of the sample evaluated?

(A) 15 kPa
(B) 30 kPa
(C) 34 kPa
(D) 60 kPa

SITUATION FOR PROBLEMS 36–40

A municipal water storage tank is to be supported by a circular raft (mat) foundation placed on the surface of the soil. The diameter of the foundation is 40 ft. The maximum load exerted on the soil when the tank is full is 2,500,000 lbf. The following data was taken from a boring log and other soil tests from the proposed site.

Elev. 0.0:
　　ground surface
　　well-graded sand and gravel
　　unit weight: 130 lbf/ft^3
　　allowable bearing pressure: 3000 lbf/ft^2

Elev. 5.0:
　　encountered GWT
　　soft, brown clay
　　unit weight: 132.4 lbf/ft^3
　　compression index: 0.34
　　unloaded, original void ratio: 1.15
　　coefficient of consolidation: 0.10 ft^2/day

Elev. 15.0:
　　encountered thick, impervious rock layer

36. What is the factor of safety in bearing?

(A) 1.2
(B) 1.5
(C) 2.3
(D) 6.0

37. What is the vertical stress at the midpoint of the clay layer prior to the installation of the water tank?

(A) 700 lbf/ft^2
(B) 1000 lbf/ft^2
(C) 1200 lbf/ft^2
(D) 1400 lbf/ft^2

38. What is the vertical stress at the midpoint of the clay layer below the center of the foundation?

(A) 1000 lbf/ft^2
(B) 2000 lbf/ft^2
(C) 3000 lbf/ft^2
(D) 4000 lbf/ft^2

39. What will be the primary settlement of the water tank?

(A) 0.10 ft or less
(B) 0.25 ft
(C) 0.50 ft
(D) 0.75 ft or more

40. How long will it take for 80% of the primary settlement to occur?

(A) 300 days
(B) 400 days
(C) 500 days
(D) 600 days

STOP!

DO NOT CONTINUE!

This concludes the Morning Session of the examination. If you finish early, check your work and make sure that you have followed all instructions. After checking your answers, you may turn in your Examination Booklet and Answer Sheet and leave the examination room. Once you leave, you will not be permitted to return to work or change your answers.

Instructions

In accordance with the rules established by your state, you may use textbooks, handbooks, bound reference materials, and any approved battery- or solar-powered, silent calculator to work this examination. However, no blank papers, writing tablets, unbound scratch paper, or loose notes are permitted. Sufficient room for scratch work is provided in the Examination Booklet.

You are not permitted to share or exchange materials with other examinees. However, the books and other resources used in this afternoon session do not have to be the same as were used in the morning session.

This portion of the examination is divided into five specialty options. You may select any one of the specialty options, regardless of your work experience. However, you may not "jump around" and solve questions from more than one specialty option.

You will have four hours in which to work this session of the examination. Your score will be determined by the number of questions that you answer correctly. There is a total of 40 questions in each specialty option. These 40 questions are divided into sets of two to ten questions, each set pertaining to a particular scenario. All 40 questions in the specialty option must be worked correctly in order to receive full credit on the exam. There are no optional questions. Each question is worth 1 point. The maximum possible score for this section of the examination is 40 points.

Partial credit is not available. No credit will be given for methodology, assumptions, or work written in your Examination Booklet.

Record all of your answers on the Answer Sheet. No credit will be given for answers marked in the Examination Booklet. Mark your answers with a no. 2 pencil. Answers marked in pen may not be graded correctly. Marks must be dark and must completely fill the bubbles. Record only one answer per question. If you mark more than one answer, you will not receive credit for the question. If you change an answer, be sure the old bubble is erased completely, as incomplete erasures may be misinterpreted as answers.

If you finish early, check your work and make sure that you have followed all instructions. After checking your answers, you may turn in your Examination Booklet and Answer Sheet and leave the examination room.

Once you leave, you will not be permitted to return to work or change your answers.

When permission has been given by your proctor, break the seal on the Examination Booklet. Check that all pages are present and legible. If any part of your Examination Booklet is missing, your proctor will issue you a new Booklet.

Do not work any questions from the Morning Session during the second four hours of this exam.

WAIT FOR PERMISSION TO BEGIN

Name: _____
       Last   First   Middle Initial

Examinee number: _____

Examination Booklet number: _____

Specialty Option to be graded: _____

Principles and Practice of Engineering Examination

Afternoon Session
Sample Examination

Specialty Options

Afternoon Session
Water Resources

A newly developed 150 ac single-family residential subdivision is drained by an intermittent straight stream with a uniform cross section. The area in flow is approximately rectangular, 4 ft wide and 6 ft deep. The stream is overgrown with weeds. The elevation change is measured as 1.54 ft over a total length of 21+35.69 sta.

KAS 0797

41. The geometric slope of the existing stream is most nearly

(A) 0.0007 ft/ft

(B) 0.001 ft/ft

(C) 0.009 ft/ft

(D) 0.07 ft/ft

42. Manning's roughness coefficient, n, for the existing stream is most nearly

(A) 0.017

(B) 0.025

(C) 0.035

(D) 0.060

43. The capacity of the existing stream when flowing full is most nearly

(A) 36 ft^3/sec

(B) 59 ft^3/sec

(C) 73 ft^3/sec

(D) 120 ft^3/sec

44. A peak post-development flow of 138 ft^3/sec is expected to leave the site. Since this exceeds the capacity of the existing stream, a trapezoidal earthen channel will be produced by excavating the sides of the existing stream. All vegetation and debris will be removed, although the depth and original base width will remain unchanged. The earthen walls will be firmly compacted.

The maximum side slope (horizontal:vertical) for the channel should be

(A) 0.5:1

(B) 1.5:1

(C) 3:1

(D) 4:1

45. A side slope of 1.5:1 [horizontal:vertical] is used with a value of Manning's roughness coefficient of 0.018. The normal depth of flow (free surface to horizontal bottom) in the trapezoidal channel when carrying the peak flow is most nearly

(A) 3.2 ft

(B) 3.7 ft

(C) 4.8 ft

(D) 5.3 ft

46. Assuming the depth of flow is 3.9 ft, the velocity of flow in the trapezoidal channel when carrying the peak flow is most nearly

(A) 2.5 ft/sec

(B) 3.0 ft/sec

(C) 3.7 ft/sec

(D) 4.9 ft/sec

47. The capacity of the trapezoidal channel flowing completely full is most nearly

(A) 230 ft^3/sec

(B) 360 ft^3/sec

(C) 410 ft^3/sec

(D) 490 ft^3/sec

48. The critical depth in the trapezoidal channel for a discharge of 138 ft^3/sec is most nearly

(A) 1.5 ft

(B) 2.5 ft

(C) 2.9 ft

(D) 3.0 ft

49. If the earthen trapezoidal channel is lined with riprap such that the area in flow does not change, the most likely effect will be

(A) increased flow depth and increased channel capacity

(B) increased flow depth and decreased channel capacity

(C) decreased flow depth and increased channel capacity

(D) decreased flow depth and decreased channel capacity

50. Manning's roughness coefficient, n, for a riprap-lined channel is most nearly

(A) 0.018

(B) 0.025

(C) 0.035

(D) 0.050

SITUATION FOR PROBLEMS 51–60

A pumping station serving a small rural community lifts 60°F water from a storage reservoir. The surface of the reservoir is 20 ft below the pump centerline. Water is discharged freely 20 ft above the pump centerline into a second reservoir. The water travels through 2825 ft of 6 in schedule-40 flanged steel pipe. The discharge line includes four 90°, long-radius elbows, two swing check valves, one gate valve, one butterfly valve, and one venturi flow meter. All valves are 100% open. Entrance and exit losses are insignificant. The pump impeller is radial-vaned. The pump manufacturer has provided the following pump performance data for cold, clear water. Assume the pump station operates as intended.

flow rate (gpm)	head (ft)
100	63 ft
200	54 ft
300	45 ft
400	40 ft
500	37 ft

51. What is the approximate equivalent length of 6 in schedule-40 flanged steel pipe?

(A) 2850 ft

(B) 2940 ft

(C) 3000 ft

(D) 3090 ft

52. What is the approximate rate of water flow through the pipe?

(A) 190 gal/min

(B) 220 gal/min

(C) 250 gal/min

(D) 280 gal/min

53. What is the approximate total head added by the pump?

(A) 45 ft

(B) 50 ft

(C) 55 ft

(D) 60 ft

54. If long sections of schedule-80 pipe are installed instead of schedule-40 pipe, the friction loss will

(A) not change, and the flow rate will remain the same.

(B) increase, and the flow rate will decrease.

(C) increase, and the flow rate will increase.

(D) decrease, and the flow rate will decrease.

55. What approximate hydraulic power does the pump develop?

(A) 1 hp

(B) 2 hp

(C) 3 hp

(D) 5 hp

56. What is the approximate minimum-sized motor required to drive the pump?

(A) 2 hp

(B) 3 hp

(C) 5 hp

(D) 7.5 hp

57. The pump is driven by a 4-pole, 60 Hz induction motor. What is the approximate specific speed of the pump?

(A) 530 rpm

(B) 980 rpm

(C) 1400 rpm

(D) 1800 rpm

58. What is the approximate pump efficiency?

(A) 55%

(B) 65%

(C) 75%

(D) 85%

59. During an emergency repair, an impeller with a diameter of 90% of the original impeller diameter is substituted. If the pump is run at the same speed, by what percentage will the flow rate decrease?

(A) 1%

(B) 3.2%

(C) 4.6%

(D) 10%

60. Which redesign would probably result in better performance?

(A) Use two identical pumps, identical to the current pump, in parallel.

(B) Use a pump with a different type of impeller.

(C) Reduce the suction line diameter.

(D) Locate the pump at a lower elevation.

SITUATION FOR PROBLEMS 61–65

Untreated water is stored in an uncovered municipal reservoir with an average winter depth of 15 ft and average winter temperature of 40°F. During the summer, the land surrounding the reservoir accumulates windblown silt (approximately spherical particles with a specific gravity of 2.65) and organic debris. The first winter storm washes all of this material into the reservoir. After the storm, the floating debris is quickly skimmed off, leaving the silt in the water to settle out. The water is considered clear enough to discharge when the largest particle in suspension has a size of 4.0×10^{-4} in. There is adequate storage elsewhere to provide treated water for 10 hr after a storm.

61. What is the minimum settling velocity of a particle based on available storage?

(A) 0.7 ft/hr

(B) 1.5 ft/hr

(C) 15 ft/hr

(D) 18 ft/hr

62. What is the settling Reynolds number of the maximum-sized particle remaining in suspension after 10 hr?

(A) 0.0008

(B) 0.3

(C) 14

(D) 1200

63. What is the expression that best describes the settling characteristics?

(A) Blasius equation

(B) Froude number

(C) Kutta-Joukowsky theorem

(D) Stokes' law

64. What is the actual maximum settling velocity of suspended particles in the water?

(A) 0.7 ft/hr

(B) 1.5 ft/hr

(C) 15 ft/hr

(D) 18 ft/hr

65. What is the maximum time needed for the reservoir water to become clear enough to be withdrawn?

(A) 6 hr

(B) 8 hr

(C) 10 hr

(D) 21 hr

SITUATION FOR PROBLEMS 66–70

A flat-bottomed concrete channel with inclined sides carries water at the rate of 2200 ft³/sec at a uniform depth of 10 ft above the bottom. The sides are constructed with a vertical:horizontal slope of 1:2, and the Manning roughness coefficient is 0.013. The elevation drops 2 ft/mi along the channel length.

66. What is the geometric slope for this channel?

(A) 0.0002 ft/ft

(B) 0.0004 ft/ft

(C) 0.08 ft/ft

(D) 2.0 ft/ft

67. What is the wetted length of each inclined side?

(A) 5 ft

(B) 15 ft

(C) 20 ft

(D) 22 ft

68. What is the width of the channel at its free surface?

(A) 37 ft

(B) 44 ft

(C) 51 ft

(D) 55 ft

69. What is the equivalent diameter of the channel?

(A) 22 ft

(B) 25 ft

(C) 29 ft

(D) 35 ft

70. What should be the vertical:horizontal slope of the channel sides to maximize the flow rate?

(A) 1:0.5

(B) 1:0.6

(C) 1:1.4

(D) 1:1.7

SITUATION FOR PROBLEMS 71-80

Water is released from a wide sluice gate into a rectangular stilling basin of the same width. The bottom of the basin is horizontal and is constructed of concrete. The average discharge velocity is 62 ft/sec, measured a short distance downstream at the point of minimum discharge depth. The minimum depth of 4.08 ft. Beyond the measurement point, the depth increases gradually. Somewhere downstream, the flow experiences a hydraulic jump to attain the depth of the stilling basin. The depth of the flow just before the hydraulic hump is 4.57 ft.

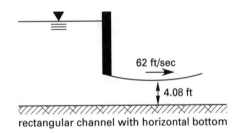

rectangular channel with horizontal bottom

71. What is the most likely value of the Manning roughness coefficient for the bottom of the stilling basin?

(A) 0.004

(B) 0.014

(C) 0.018

(D) 0.022

72. The reason the flow initially decreases in depth is most likely

(A) gravitational attraction between the earth and the water

(B) frictional retarding between the water and the stilling basin floor

(C) the proximity of the flow to the critical condition

(D) the geometry of the sluice gate opening

73. At the point of minimum depth, the flow is

(A) critical

(B) subcritical

(C) accelerated

(D) retarded

74. What is the hydraulic radius of the flow at the point of minimum depth?

(A) 0.45 ft

(B) 1.2 ft

(C) 2.9 ft

(D) 4.1 ft

75. What is the approximate velocity just before the hydraulic jump?

(A) 51 ft/sec

(B) 55 ft/sec

(C) 57 ft/sec

(D) 61 ft/sec

76. What is the approximate depth of the stilling basin after the hydraulic jump?

(A) 6.3 ft

(B) 7.9 ft

(C) 25 ft

(D) 27 ft

77. If the depth after the hydraulic jump is 25 ft, what is the approximate velocity just after the hydraulic jump?

(A) 9.3 ft/sec

(B) 9.7 ft/sec

(C) 10.1 ft/sec

(D) 14.7 ft/sec

78. What is the average energy gradient between the locations of the minimum depth and the hydraulic jump?

(A) 0.0

(B) 0.037

(C) 0.052

(D) 0.064

79. What is the approximate distance between the point of minimum depth and the hydraulic jump?

(A) 210 ft

(B) 260 ft

(C) 320 ft

(D) 350 ft

80. The specific energy loss per unit width in the hydraulic jump is most nearly

(A) 14 ft

(B) 18 ft

(C) 23 ft

(D) 27 ft

STOP!

DO NOT CONTINUE!

This concludes the Afternoon Session of the examination. If you finish early, check your work and make sure that you have followed all instructions. After checking your answers, you may turn in your Examination Booklet and Answer Sheet and leave the examination room. Once you leave, you will not be permitted to return to work or change your answers.

Afternoon Session
Structural

SITUATION FOR PROBLEMS 81–90

An 18 ft long reinforced concrete cantilever beam is 18 in wide. The beam is not exposed to earth or weather. In addition to its own weight, the beam supports a uniformly distributed dead load of 1500 lbf/ft along its cantilevered length. The beam also supports a vertical live load of 10,000 lbf at its free end. The concrete compressive strength is 3000 psi. The yield strength of the steel bars is 60 ksi. No. 9 tensile rebar, no. 3 stirrups, and strength design have been specified.

81. What is the approximate ultimate moment acting at the built-in end of the beam?

(A) 250,000 ft-lbf or less

(B) 500,000 ft-lbf

(C) 800,000 ft-lbf

(D) 1,000,000 ft-lbf or more

82. The minimum reinforcement ratio is closest to

(A) 0.001

(B) 0.003

(C) 0.005

(D) 0.007

83. The reinforcement ratio selected is 50% of the maximum permitted. The reinforcement ratio is most nearly

(A) 0.008

(B) 0.010

(C) 0.016

(D) 0.020

84. The final beam depth (measured to the centroid of the tensile steel) after correcting for initial assumptions is closest to

(A) 30 in or less

(B) 34 in

(C) 37 in

(D) 40 in or more

85. How many no. 9 bars are required?

(A) 6

(B) 7

(C) 8

(D) 9

86. What minimum percentage of the tensile steel should be located above the neutral axis?

(A) 0%

(B) 10%

(C) 20%

(D) 100%

87. Assume that the beam depth (measured to the centroid of the tensile steel) is 37 in. What is the minimum beam height (measured from the top to the bottom surface)?

(A) 38.5 in or less

(B) 39.0 in

(C) 39.5 in

(D) 40.0 in or more

88. The cracked moment of inertia is closest to

(A) 39,000 in^4

(B) 41,000 in^4

(C) 43,000 in^4

(D) 45,000 in^4

89. The effective moment of inertia is closed to

(A) 42,000 in^4

(B) 44,000 in^4

(C) 46,000 in^4

(D) 48,000 in^4

90. The instantaneous deflection at the cantilevered end due to the dead and live load is most nearly

(A) 0.4 in

(B) 0.6 in

(C) 0.8 in

(D) 1.0 in

SITUATION FOR PROBLEMS 91–100

A tilt-up building 30 ft wide has a flat roof constructed of plywood and covered with hot-mop asphalt waterproofing. Including an allowance for roof-mounted air conditioning equipment, the dead load on the roof is 20 lbf/ft^2. The live load has been estimated as 25 lbf/ft^2 for design purposes. Straight glued-laminated (glulam) beams are placed across the building width every 10 ft (i.e., on 10 ft centers) to support the roof. The beams are installed such that the laminations are horizontal. There are no interior columns to support the beams. The beams were produced from western species Douglas fir 2×6's without any special tensile reinforcement laminations and are marked "24F-V4." It has been decided not to evaluate lateral buckling of the glulams, and that duration of load factors can be neglected in the stress calculations. The beams are considered to be simply supported.

91. Approximately what total load must each interior glulam beam support?

   (A) 7500 lbf or less

   (B) 8500 lbf

   (C) 13,500 lbf

   (D) 14,000 lbf or more

92. What does the designation "24F" indicate?

   (A) The maximum extreme fiber tensile stress is 2400 psi.

   (B) The beam is suitable for use with spacings up to 24 ft without considering lateral bracing.

   (C) The modulus of elasticity of the beam is 24×10^6 psi.

   (D) The beam has a 24 hr fire rating.

93. What does the designation "V4" indicate?

   (A) The maximum shear stress is 400 psi.

   (B) The glulam was assembled from visually inspected boards.

   (C) The glulam is architectural grade, free of visual blemishes.

   (D) The glulam is suitable only for interior use.

94. What is the approximate width of each glulam?

   (A) $5\frac{1}{8}$ in

   (B) $5\frac{3}{8}$ in

   (C) $5\frac{5}{8}$ in

   (D) $5\frac{7}{8}$ in

95. What is the most likely reason for not considering lateral buckling?

   (A) Glulams are extremely stiff and have inherent resistance against buckling.

   (B) Continuous connection to the roof provides the needed lateral support.

   (C) The glulam's designation permits disregarding lateral buckling in specific installations.

   (D) Simply supported ends rotate before lateral buckling occurs.

96. Assume that the total load carried by each glulam beam is 14,250 lbf and that the roof equipment is evenly spaced around the roof. What is the minimum number of glue laminations each glulam should have?

   (A) 10

   (B) 11

   (C) 12

   (D) 14

97. What is the approximate maximum permitted compressive stress perpendicular to the grain?

   (A) 400 psi

   (B) 650 psi

   (C) 1200 psi

   (D) 2400 psi

98. If a beam is accidentally installed upside down (i.e., camber-down), what would be the allowable tensile stress in the lower part of the beam (i.e., the part closest to be ground)?

   (A) 650 psi

   (B) 1200 psi

   (C) 1850 psi

   (D) 2400 psi

99. If the beam is installed properly (i.e., camber-up), what is the approximate allowable shear stress parallel to the grain in the glulam?

   (A) 180 psi

   (B) 190 psi

   (C) 210 psi

   (D) 240 psi

100. What is the approximate camber required at the midpoint of an interior glulam in order to appear horizontal from below when fully loaded?

(A) 0.5 in

(B) 1.0 in

(C) 1.5 in

(D) 2.0 in

SITUATION FOR PROBLEMS 101–110

A through-bridge spanning a small creek is 20 ft wide and 90 ft long. The deck has a negligible dead weight. The bridge carries a uniform live load of 20 lbf/ft^2 over the entire deck. The deck is supported by two sets of two trusses connected by a hinge at the center of each side. Both trusses are constructed of A36 steel members with non-moment-resisting welded connections. The bridge is loaded at the lower panel points by the floor beams. No part of the deck load is transmitted to the central hinge point. All truss member loads are axial. All truss member weights are negligible.

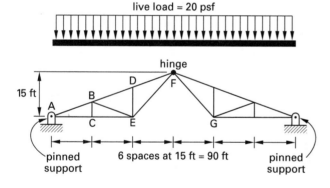

101. The vertical reaction at point A on one of the trusses is closest to

(A) 4.5 kips

(B) 9.0 kips

(C) 18 kips

(D) 36 kips

102. The vertical load transmitted to one of the trusses at point C is most nearly

(A) 1.5 kips

(B) 3.0 kips

(C) 4.5 kips

(D) 6.0 kips

103. What is the approximate vertical load that is transmitted to one of the trusses at point E?

(A) 1.5 kips

(B) 3.0 kips

(C) 4.5 kips

(D) 6.0 kips

104. What is the horizontal reaction at point A for one of the trusses?

(A) 0 kips

(B) 3 kips

(C) 6 kips

(D) 12 kips

105. What is the approximate load carried by member AB?

(A) 4.7 kips

(B) 9.2 kips

(C) 16 kips

(D) 24 kips

106. What is the approximate load carried by member DE?

(A) 0 kips

(B) 3.0 kips

(C) 4.7 kips

(D) 8.5 kips

107. What is the approximate load carried by member AC?

(A) 3.0 kips

(B) 3.7 kips

(C) 11 kips

(D) 19 kips

108. The allowable stress in the tensile members is closest to

(A) 18 ksi

(B) 19 ksi

(C) 22 ksi

(D) 36 ksi

109. Which shape in the following list is the most economical W8 beam for member AC?

(A) W 8 × 10

(B) W 8 × 13

(C) W 8 × 18

(D) W 8 × 24

110. Which shape in the following list is the most economical W8 beam for member AB?

(A) W8 × 10

(B) W8 × 18

(C) W8 × 24

(D) W8 × 31

SITUATION FOR PROBLEMS 111–120

A concrete mix is based on the following table.

material	specific gravity	mass (lbm)	volume (ft^3)
cement	3.15	517	??
fine aggregate	2.65	??	10.80
coarse aggregate	??	2679	16.00
water	1.00	232	??

111. The volume of cement is most nearly

(A) 2.5 ft^3

(B) 2.6 ft^3

(C) 2.7 ft^3

(D) 2.8 ft^3

112. The mass of fine aggregate is most nearly

(A) 1600 lbm

(B) 1700 lbm

(C) 1750 lbm

(D) 1800 lbm

113. The specific gravity of the coarse aggregate is most nearly

(A) 2.60

(B) 2.65

(C) 2.68

(D) 2.75

114. The total volume of the mix is most nearly

(A) 29 ft^3

(B) 31 ft^3

(C) 33 ft^3

(D) 35 ft^3

115. Scaling of concrete surfaces is usually the result of

(A) high cement content

(B) high chloride content

(C) aggregate-alkali reaction

(D) low air entrainment or poor finishing

116. A 6 in by 12 in concrete cylinder cast in the field is subjected to 100,000 lbf compressive load. This cylinder will, most likely, be used to determine a minimum

(A) 400 psi splitting tensile strength

(B) 1000 psi modulus of rupture

(C) 1400 psi compressive strength

(D) 3500 psi compressive strength

117. The maximum amount of an accelerator in a mixture is 2% of the cement weight. The maximum mass of this accelerator in 8 yd^3 of a six-bag mix is

(A) 80 lbm

(B) 90 lbm

(C) 100 lbm

(D) 110 lbm

118. The type of cement used for high early-strength concrete is

(A) type I

(B) type II

(C) type III

(D) type IV

119. The term "SSD" refers to

(A) supersaturated dry

(B) soaked saturated dry

(C) saturated surface dry

(D) submerged saturated dry

120. Which of the following represents a pozzolan with a low calcium oxide and high carbon content?

(A) class C flyash

(B) class F flyash

(C) silica fume

(D) bentonite clay

STOP!

DO NOT CONTINUE!

This concludes the Afternoon Session of the examination. If you finish early, check your work and make sure that you have followed all instructions. After checking your answers, you may turn in your Examination Booklet and Answer Sheet and leave the examination room. Once you leave, you will not be permitted to return to work or change your answers.

Afternoon Session
Transportation

BAS 1296

SITUATION FOR PROBLEMS 121–130

The circular curve shown represents the centerline of a two-lane rural highway that passes through level surroundings. The highway has two 12 ft lanes with 6 ft shoulders. The arc basis degree of curvature is 2°, and the design speed is 40 mph. The point of intersection is located at 423968.68 N, 268236.42 E. The curve is in the vicinity of an established Civil War cemetery.

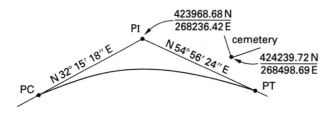

121. The coordinates of the PT are most nearly

(A) 424298.78 N
    268706.80 E

(B) 424297.26 N
    268715.97 E

(C) 424309.47 N
    268707.33 E

(D) 424897.66 N
    268136.44 E

122. The bearing of the radius line meeting the curve at the PT is most nearly

(A) N 35°3′36″ E

(B) N 54°56′24″ E

(C) S 34°2′16″ E

(D) S 35°3′36″ E

123. The coordinates of the curve's center (i.e., the origin of the radius line) are most nearly

(A) 421965.80 N
    270358.01 E

(B) 421953.80 N
    270352.43 E

(C) 421954.02 N
    270341.97 E

(D) 421965.80 N
    270356.43 E

124. The coordinates of the PC are most nearly

(A) 423482.71 N
    267980.45 E

(B) 423507.68 N
    267929.74 E

(C) 423482.71 N
    267929.74 E

(D) 423661.99 N
    267750.45 E

125. Assume that the point on the curve closest to the corner of the cemetery has coordinates 424180.59 N, 268549.70 E. What is the distance from the curve to the corner of the cemetery?

(A) 77.5 ft

(B) 77.8 ft

(C) 78.1 ft

(D) 78.4 ft

126. What is the approximate angle between the radius passing through the PT and the radius passing through the corner of the cemetery?

(A) 2°30′00″

(B) 3°58′24″

(C) 3°58′48″

(D) 4°57′32″

127. What would be a reasonable recommendation for superelevation on this curve?

(A) 0.020

(B) 0.030

(C) 0.040

(D) 0.080

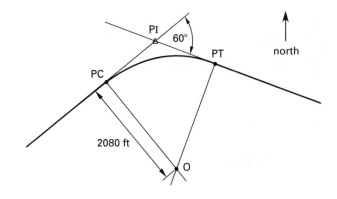

128. What would be a reasonable recommendation for a spiral runoff length (not including tangent runout) for this curve?

(A) 100 ft

(B) 125 ft

(C) 150 ft

(D) 175 ft

129. What minimum clearance (i.e., "clear zone") distance from the edge of the traveled pavement to obstructions should generally be provided along the traveled way?

(A) 2 ft or less

(B) 4 ft

(C) 6 ft

(D) 10 ft or more

curve radius, R (ft)	superelevation, e	runoff in the curve, L (ft)	
		2 lanes	4 lanes
22,920	NC	0	0
11,460	NC	0	0
7640	RC	0	0
5730	0.020	150	150
3820	0.028	150	150
2865	0.035	150	150
2290	0.040	150	150
1910	0.045	150	160
1640	0.048	150	170
1430	0.052	150	180
1145	0.056	150	200
955	0.059	150	210

130. What is the approximate minimum stopping sight distance for this curve?

(A) 200 ft

(B) 275 ft

(C) 450 ft

(D) 475 ft

131. The degree of the curve is most nearly

(A) 2°00′

(B) 2°45′

(C) 2°76′

(D) 3°16′

132. The station of the PI is most nearly

(A) 24+40

(B) 22+20

(C) 34+20

(D) 48+20

SITUATION FOR PROBLEMS 131–140

The centerline alignment of a circular curve in a two-lane roadway is shown. Each lane is 12 ft wide. There is no shoulder. The PC station is 12+40. The curve radius is 2080 ft. The interior angle is 60°. The roadway is to be superelevated around the curve. The axis of rotation will be the centerline. The criteria for the superelevation rate and runoff length are given in the table. Tangent runouts are twice the runoff lengths listed.

MH 0000

133. The station of the PT is most nearly

(A) 24+40

(B) 22+20

(C) 34+20

(D) 48+20

134. The length of this curve's long chord is most nearly

(A) 1000 ft

(B) 1200 ft

(C) 2100 ft

(D) 2200 ft

135. The station for the beginning of superelevation transition is most nearly

(A) 9+40

(B) 10+90

(C) 11+40

(D) 11+90

136. The maximum superelevation is most nearly

(A) 0.020

(B) 0.040

(C) 0.043

(D) 0.059

137. Assuming a superelevation of 0.050, a linear (constant) superelevation transition, and a normal crown cross slope of 2%, the station at which the reverse crown is achieved is most nearly

(A) 9+40

(B) 11+97

(C) 12+40

(D) 13+90

138. The roadway profile is on a constant uphill grade of 0.750%. The elevation of the centerline at the PC is 170 ft. Assume the superelevation is 0.048. The elevation of the outside (higher) edge of the roadway at the midpoint along the curve is most nearly

(A) 177.59 ft

(B) 178.17 ft

(C) 178.41 ft

(D) 178.74 ft

139. A proposed realignment of the roadway calls for offsetting the alignment beyond the PT by 50 ft, parallel to and to the northeast of the existing alignment. The PC is to remain at the same location with the same stationing. The station of the proposed PT is most nearly

(A) 34+18

(B) 34+20

(C) 34+68

(D) 35+23

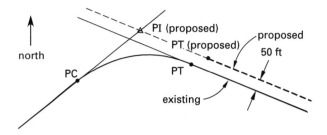

140. The station of the proposed PI is most nearly

(A) 23+30

(B) 24+82

(C) 24+41

(D) 24+99

SITUATION FOR PROBLEMS 141–150

A six-lane freeway with an ideal 70 mph free-flow speed passes through rolling terrain. There is an interchange approximately every 2 mi. A few sections have grades greater than 3%, but none of these sections are longer than $1/4$ mile in length. Each lane is 10 ft wide. The minimum clear distance between overpass abutments, curve rails, and other roadside obstructions is 2 ft on both shoulders. A recent traffic study for a particular portion of the daily commute period shows that the directional weekday volume is 2400 vehicles per hour in one direction. An average of 750 vehicles pass by during the busiest 15 min. The traffic stream consists of 14% trucks, 8% buses, and 4% recreational vehicles.

141. How many lanes carry traffic in each direction?

(A) 1

(B) 2

(C) 3

(D) 6

142. What is the approximate peak hour factor?

(A) 0.6

(B) 0.7

(C) 0.8

(D) 0.9

143. What is the approximate heavy vehicle factor?

(A) 0.6

(B) 0.7

(C) 0.8

(D) 0.9

144. What is the 15 min passenger-car equivalent flow rate?

(A) 800 pcphpl

(B) 1100 pcphpl

(C) 1300 pcphpl

(D) 1500 pcphpl

145. What is the approximate volume-to-capacity ratio?

(A) 0.50

(B) 0.60

(C) 0.70

(D) 0.80

146. What is the approximate free-flow speed?

(A) 59 mph

(B) 61 mph

(C) 63 mph

(D) 65 mph

147. Assuming the free-flow speed was 60 mph and all other values are as determined, at which level of service (LOS) is the freeway operating?

(A) level of service B

(B) level of service C

(C) level of service D

(D) level of service E

148. Assuming the free-flow speed was 60 mph and all other values are as determined, what would the volume-to-capacity ratio have to be in order to improve the (capacity) level of service to the next lower (better) level of service (LOS)?

(A) 0.30

(B) 0.40

(C) 0.60

(D) 0.80

149. Assuming a 60 mph free-flow speed and no other changes, what minimum number of lanes (one direction) is needed to provide LOS B?

(A) 3

(B) 4

(C) 5

(D) 6

150. Assume the free-flow speed is 60 mph and no other factors change. Traffic volumes are growing at the rate of 5% per year. An additional lane in each direction will be added when the LOS reaches D. Although the right-of-way has already been purchased, it will take 2 years to publish the job, select contractors, and construct the additional lanes. When should the request for bids be published?

(A) immediately

(B) 1 year from now

(C) 2 years from now

(D) 3 years from now

SITUATION FOR PROBLEMS 151–160

The approach tangent to an equal-tangent vertical curve has a slope of +3%. The slope of the departure tangent is −2%. These two tangents intersect at sta 26+00 and elev 231.00 ft. A set of subway rails passes below and perpendicular to the curve at sta 28+50 on the departure side. The maximum elevation of the railbed at that point is 195.00 ft. Allowing for depth of the roadway and the overpass structure, the minimum clearance between the railbed and the roadway surface is 26 ft.

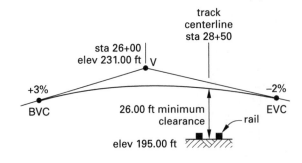

151. The horizontal distance from the BVC of any point on the curve is

(A) proportional to the square root of the vertical distance from the curve

(B) inversely proportional to the square root of the vertical distance from the curve

(C) proportional to the square of the vertical distance from the curve

(D) inversely proportional to the square of the vertical distance from the curve

152. What is the approximate minimum permissible roadway elevation at that point where the subway line crosses?

(A) 170 ft

(B) 180 ft

(C) 200 ft

(D) 220 ft

153. What is the approximate longest possible vertical curve that can be constructed and still provide the required clearance?

(A) 14.7 sta

(B) 15.9 sta

(C) 16.5 sta

(D) 17.3 sta

154. What is the approximate elevation of the BVC?

(A) 196 ft

(B) 205 ft

(C) 206 ft

(D) 207 ft

155. What is the approximate station of the BVC?

(A) sta 17+20

(B) sta 17+80

(C) sta 18+00

(D) sta 18+70

156. What is the approximate station of the EVC?

(A) sta 27+60

(B) sta 31+70

(C) sta 33+10

(D) sta 34+20

157. What is the approximate elevation of EVC?

(A) 214.1 ft

(B) 214.5 ft

(C) 214.9 ft

(D) 215.3 ft

158. What is the approximate station of the highest point on the curve?

(A) sta 24+20

(B) sta 26+00

(C) sta 26+50

(D) sta 27+70

159. What is the approximate elevation of the highest point on the curve?

(A) 221.1 ft

(B) 221.6 ft

(C) 222.3 ft

(D) 223.9 ft

160. Which of the following equations best describes points on the vertical curve? y is in feet; x is in stations, measured from the BVC.

(A) $y = 205 + 2x + 0.15x^2$

(B) $y = 206 - 3x - 0.30x^2$

(C) $y = 206 + 3x - 0.15x^2$

(D) $y = 209 - 2x + 0.30x^2$

STOP!

DO NOT CONTINUE!

This concludes the Afternoon Session of the examination. If you finish early, check your work and make sure that you have followed all instructions. After checking your answers, you may turn in your Examination Booklet and Answer Sheet and leave the examination room. Once you leave, you will not be permitted to return to work or change your answers.

Afternoon Session
Environmental

A small town with 4000 inhabitants generates wastewater at the average rate of 0.32 MGD. The wastewater enters the plant with a BOD_5 of 235 mg/L. Stabilization ponds are used as the town's main treatment method. Multiple ponds are used in various series and parallel configurations. During the winter, discharge from the ponds is blocked, and the pond depths are permitted to increase from 2 ft to 5 ft. The average evaporation loss from the pond during the winter is 0.08 in/day.

161. If the combined total acreage in use is 30 ac, what is the BOD loading on the combined pond area?

- (A) 10 lbm/ac-day
- (B) 20 lbm/ac-day
- (C) 40 lbm/ac-day
- (D) 80 lbm/ac-day

162. If the actual domestic BOD loading is 35 lbm/ac-day when the ponds are used in one configuration, how many people are supported per acre?

- (A) 100 people/ac
- (B) 200 people/ac
- (C) 400 people/ac
- (D) 800 people/ac

163. If 30 ac are in use at the beginning and end of the winter, what is the duration of the winter season?

- (A) 60 days
- (B) 80 days
- (C) 100 days
- (D) 120 days

164. The town is entertaining a request from an industrial plating facility. The plating facility wants to discharge its plating solutions directly into the ponds. The plating facility is willing to construct a separate, monitored, dedicated discharge line directly from its facility to the ponds. What factors should be considered?

   I.   the ability of the pond to process the plating waste

   II.   the construction details of the ponds and dikes

   III.   a decrease in process efficiency due to cloudiness caused by added suspended solids

   IV.   an increase in pond vegetation, flies, and worms

   V.   infiltration of facility waste into the town's water mains

- (A) I and II
- (B) I, II, and III
- (C) II and IV
- (D) I, II, III, IV, and V

165. The town is entertaining a request from an industrial meat packing facility. The meat packing facility wants to discharge its processing and cleaning water directly into the ponds. In keeping with USDA standards, except for chlorine disinfectant, no chemicals of any kind are used by the facility. The meat packing facility is willing to construct a separate, monitored, dedicated discharge line directly from its facility to the ponds. What factors should be considered?

   I.   the ability of the pond to process the plating waste

   II.   the construction details of the ponds and dikes

   III.   a decrease in process efficiency due to cloudiness caused by added suspended solids

   IV.   an increase in pond vegetation, flies, and worms

   V.   infiltration of facility waste into the town's water mains

- (A) II and III
- (B) I, II, and III
- (C) III and IV
- (D) I, II, III, IV, and V

SITUATION FOR PROBLEMS 166–170

A wastewater treatment plant receives 1 MGD from a town with a population of 13,500 people. The plain sedimentation process removes 100 mg/L of BOD_5. The discharge from the plain sedimentation process has a BOD_5 of 175 mg/L and contains essentially no dissolved oxygen. After passing through an activated sludge process, the plant effluent BOD_5 is 65 mg/L and the dissolved oxygen content is 3.6 mg/L. Aerators supply air in the secondary processing at the rate of 0.005 ft^3-L/gal-mg (i.e., 0.005 cubic feet per gallon of sewage per milligram per liter of BOD_5 removed).

166. What is the per-capita population equivalent?

- (A) 0.13 lbm BOD_5/capita-day
- (B) 0.15 lbm BOD_5/capita-day
- (C) 0.17 lbm BOD_5/capita-day
- (D) 0.21 lbm BOD_5/capita-day

167. What is the approximate ideal mass of oxygen transferred to the water each day by the aerator?

- (A) 540 lbm/day
- (B) 950 lbm/day
- (C) 1500 lbm/day
- (D) 2300 lbm/day

168. Assume that the aerators supply 1000 lbm of oxygen per day. At normal conditions, the ideal volume of air provided is most nearly

- (A) 10 ft^3/min
- (B) 20 ft^3/min
- (C) 30 ft^3/min
- (D) 40 ft^3/min

169. What is the actual volume of air supplied?

- (A) 400 ft^3/min
- (B) 700 ft^3/min
- (C) 1300 ft^3/min
- (D) 3100 ft^3/min

170. Assume that the aerators supply 1000 lbm of oxygen per day. What is the approximate oxygen transfer efficiency?

- (A) 2%
- (B) 5%
- (C) 10%
- (D) 15%

SITUATION FOR PROBLEMS 171–180

The bottom of a rectangular wastewater treatment lagoon is two times as long as it is wide. The bottom area is 0.75 ac. The elevation of the bottom is 88.25 ft. The elevation at the top of the dike is 96.25 ft. The dikes are symmetrical and have a 10 ft wide driving area around the lagoon. There is a variable level effluent structure so that the lagoon may be operated as a conventional lagoon or as an aerated lagoon.

171. What approximate surface elevation would typically be applicable to a conventional lagoon?

- (A) 92 ft
- (B) 94 ft
- (C) 96 ft
- (D) 98 ft

172. What approximate surface elevation would typically be applicable to an aerated lagoon?

- (A) 92 ft
- (B) 94 ft
- (C) 96 ft
- (D) 98 ft

173. What minimum (horizontal:vertical) slope should be used on the dikes?

- (A) 2:1
- (B) 2.5:1
- (C) 3:1
- (D) 4:1

174. Assuming that the dike slope is 3:1 (horizontal: vertical), what is the surface area of the lagoon when the surface elevation is 94.25 ft?

- (A) 0.90 ac
- (B) 0.95 ac
- (C) 1.00 ac
- (D) 1.10 ac

175. Assuming that the dike slope is 3:1 (horizontal: vertical), approximately how much wastewater can the lagoon hold if it is operated with a depth of 4 ft?

- (A) 1,000,000 gal
- (B) 1,100,000 gal
- (C) 1,200,000 gal
- (D) 1,500,000 gal

176. If the detention time is 60 days, what is the treatment capacity of this lagoon?

(A) 0.015 MGD

(B) 0.017 MGD

(C) 0.019 MGD

(D) 0.025 MGD

177. Assume the surface area of the lagoon is 42,400 ft^2 and the SLR is 40 lbm BOD_5 per acre. If it is operated as a conventional lagoon, what is the population equivalent design capacity?

(A) 60 people

(B) 120 people

(C) 230 people

(D) 500 people

178. The lagoon is ultimately converted to an aerated lagoon with a flow of 90,000 gal/day. If the lagoon treats influent with a BOD_5 of 195 lbm/day with an efficiency of 89%, what is the BOD_5 of the effluent?

(A) 30 mg/L

(B) 80 mg/L

(C) 170 mg/L

(D) 230 mg/L

179. A suspended solids test is run on 50 mL of the effluent. The tare mass of the filter was 0.0985 g. The dry, filtered mass was 0.1133 g. What is the concentration of total suspended solids?

(A) 70 mg/L

(B) 150 mg/L

(C) 210 mg/L

(D) 300 mg/L

180. The filter from the previous problem was placed in a furnace. The mass of the filter ash was 0.1074 g. What percentage of the effluent total solids concentration is volatile?

(A) 20%

(B) 30%

(C) 40%

(D) 50%

SITUATION FOR PROBLEMS 181–190

High sugar-content food processing waste is to be treated in a completed-mixed, high-rate anaerobic digestion process. A portion of the methane produced is to be burned to heat the incoming waste stream. Compared to an ideal waste-to-methane process, only 70% of the theoretical methane is produced. The waste has the following characteristics.

incoming sludge temperature	65°F
flow rate	0.2 ft^3/sec
BOD, ultimate	8900 mg/L
COD	9300 mg/L
yield constant (cell tissue mass per mass ultimate BOD entering)	0.06 g/g

181. Why is the digestion process considered to be anaerobic?

(A) Anaerobic microorganisms deplete the waste of oxygen during the digestion process.

(B) The waste is not aerated during digestion.

(C) Compounds with free radicals are added to the waste in order to eliminate pipe corrosion.

(D) The digestion occurs in a partial vacuum.

182. The digestion is managed so that the waste can be processed in a minimum amount of time. Estimate the mean residence digestion time.

(A) 10 days

(B) 20 days

(C) 40 days

(D) 60 days

183. What is the approximate minimum total digester volume required?

(A) 50,000 ft^3 or less

(B) 100,000 ft^3

(C) 200,000 ft^3

(D) 300,000 ft^3 or more

184. Approximately how many digestion tanks of average dimensions should be used?

(A) 1

(B) 2

(C) 1 or 2

(D) 3 or 4

185. Approximately how much cell tissue will be produced per day?

(A) 0.25 kg/day

(B) 25 kg/day

(C) 250 kg/day

(D) 2500 kg/day

186. What is the approximate gross volume of methane gas produced each day?

(A) 100 m^3/day

(B) 1000 m^3/day

(C) 10,000 m^3/day

(D) 100,000 m^3/day

187. What is the approximate lower heating value of dry methane gas?

(A) 900 Btu/ft^3

(B) 1200 Btu/ft^3

(C) 1500 Btu/ft^3

(D) 1800 Btu/ft^3

188. At what temperature should the digester be maintained during normal operation?

(A) below 100°F

(B) 100 to 120°F

(C) 120 to 140°F

(D) above 140°F

189. What approximate volume per day of methane is required to heat the incoming waste to the optimum digestion temperature?

(A) 10,000 ft^3

(B) 20,000 ft^3

(C) 40,000 ft^3

(D) 100,000 ft^3

190. Which of the following actions will improve the efficiency of methane production?

(A) Closely monitor the acidity level.

(B) Keep the digestion temperature as low as possible.

(C) Remove aerobic bacteria by preheating the waste to 180°F.

(D) Keep the alkalinity as low as possible.

SITUATION FOR PROBLEMS 191–200

A solid waste facility recovers salable waste metals, glass, and other materials from municipal solid waste. The remaining solid waste is incinerated. The combustion heat is used to generate steam for turbines which, in turn, drives electrical generators.

The electrical generation plant consists of three power modules, each containing a furnace, a steam generator, a turbine, and an electrical generator. Each power module burns 180 tons of solid waste and 190 tons of sludge each day, seven days a week. Because of scheduled maintenance and repairs, however, the facility's utilization is 85%. No solid waste or sludge is delivered to the facility when it is down for maintenance or repairs.

When received, the solid waste contains 5% moisture and 15% material that is removed by an air classifier. Both the moisture and the classified material are removed prior to combustion.

The sludge is the by-product of wastewater treatment from mixed primary settling and trickling-filter processing.

The facility initially costs $11 million to construct. It is expected to be operational for 20 years. The annual effective interest rate for economic comparisons is 8%. Annual costs are $600,000 for labor, $700,000 for maintenance, and $100,000 for insurance.

Each power module produces 4000 kW (net) of electricity, which can be sold for $0.015 per kW-hr. The wastewater plant delivering the sludge pays $50 per ton of dry sludge solids incinerated. The values of the classified materials are as follows.

material	percent of solid waste	material value ($/ton)
sand and ash	15%	$2
glass	5%	$6
ferrous metal	5%	$50
aluminum	0.3%	$350
other metals	0.3%	$400

191. What is the approximate total amount of solid waste received by the facility during each day of operation?

(A) 180 tons

(B) 230 tons

(C) 430 tons

(D) 680 tons

192. What is the approximate amount of sludge received by the facility during each year?

(A) 60,000 tons

(B) 95,000 tons

(C) 180,000 tons

(D) 220,000 tons

193. What is the approximate size of the population generating the solid waste received by the facility?

(A) 50,000

(B) 100,000

(C) 200,000

(D) 500,000

194. Approximately how many tons of dry sludge solids are produced each day?

(A) 3 tons

(B) 15 tons

(C) 90 tons

(D) 150 tons

195. The percentage of solids contained in the sludge is closest to

(A) 1%

(B) 3%

(C) 5%

(D) 7%

196. What is the approximate equivalent uniform annualized cost of construction per ton of solid waste processed during the year over the life of the facility?

(A) $1.70/ton

(B) $2.90/ton

(C) $4.50/ton

(D) $5.40/ton

197. What is the approximate total equivalent uniform annual total cost of operation per ton of solid waste processed during the year?

(A) $12/ton

(B) $15/ton

(C) $18/ton

(D) $21/ton

198. What is the approximate value of the electricity generated per ton of solid waste processed?

(A) $2.10/ton

(B) $6.40/ton

(C) $7.50/ton

(D) $9.90/ton

199. What is the approximate total income generated per ton of waste processed?

(A) $5.40/ton

(B) $9.20/ton

(C) $13.00/ton

(D) $18.00/ton

200. What is the approximate net disposal profit (or loss) per ton of solid waste processed?

(A) $0.30/ton loss

(B) $1.00/ton profit

(C) $1.90/ton profit

(D) $2.50/ton profit

STOP!

DO NOT CONTINUE!

This concludes the Afternoon Session of the examination. If you finish early, check your work and make sure that you have followed all instructions. After checking your answers, you may turn in your Examination Booklet and Answer Sheet and leave the examination room. Once you leave, you will not be permitted to return to work or change your answers.

Afternoon Session
Geotechnical

SITUATION FOR PROBLEMS 201–204

A building site consists of impervious bedrock overlain by 10 ft of clayey soil. A 2 in thick sample of the soil was compressed in double-drainage. During the test, 50% of the ultimate consolidation occurred in the first 20 min.

201. What is the approximate coefficient of primary consolidation?

(A) 0.0034 ft^2/hr

(B) 0.0041 ft^2/hr

(C) 0.0098 ft^2/hr

(D) 0.196 ft^2/hr

202. How do graphs of void ratio (on the vertical axis) versus time (on the horizontal axis) for clayey and sandy soils compare for primary consolidation?

(A) The slope for clay is positive and more positive than the slope for sand.

(B) The slope for clay is positive and less positive than the slope for sand.

(C) The slope for clay is negative and more negative than the slope for sand.

(D) The slope for clay is negative and less negative than the slope for sand.

203. How do graphs of void ratio (on the vertical axis) versus time (on the horizontal axis) for clayey and sandy soils compare for secondary consolidation?

(A) The second derivative of the line for clay is positive and more positive than the second derivative of the line for sand.

(B) The second derivative of the line for clay is positive and less positive than the second derivative of the line for sand.

(C) The second derivative of the line for clay is negative and more negative than the second derivative of the line for sand.

(D) The second derivative of the line for clay is negative and less negative than the second derivative of the line for sand.

204. How long will it take the 10 ft thick clay layer to achieve 90% of its ultimate primary consolidation?

(A) 1.1 yr

(B) 1.7 yr

(C) 2.4 yr

(D) 4.3 yr

SITUATION FOR PROBLEM 205

A 400 ft wide concrete dam is located 190 ft above an impervious rock stratum, as shown. The hydraulic head behind the dam is 40 ft. There is no hydraulic head due to tail water. The soil beneath the dam has a coefficient of permeability of 0.005 gal/day-ft^2. In order to reduce seepage beneath the clay, a 25 ft deep, continuous sheet-pile cutoff wall has been installed.

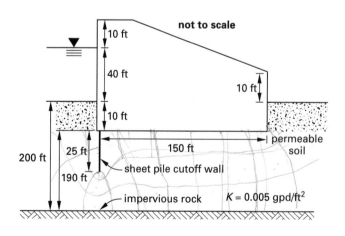

205. What is the approximate seepage rate per foot of dam width?

(A) 0.005 gal/day-ft or less

(B) 0.02 gal/day-ft

(C) 0.06 gal/day-ft

(D) 0.12 gal/day-ft or more

SITUATION FOR PROBLEMS 206–210

A Proctor test is performed on four samples. The mold volume and mass are $1/30$ ft^3 and 4200 g for each sample. The following data are collected.

sample	mass of mold and soil	water content
1	6100 g	8.2%
2	6300 g	10.1%
3	6425 g	11.7%
4	6330 g	14.8%

206. The wet mass density for sample 4 is closest to

(A) 141 lbm/ft^3

(B) 143 lbm/ft^3

(C) 145 lbm/ft^3

(D) 146 lbm/ft^3

207. The dry mass density for sample 1 is closest to

(A) 103 lbm/ft^3

(B) 116 lbm/ft^3

(C) 122 lbm/ft^3

(D) 125 lbm/ft^3

208. The relative compaction of sample 2 is approximately

(A) 90% or less

(B) 92%

(C) 94%

(D) 95% or more

209. Soils prone to drastic volume changes have which one of the following properties?

(A) liquid limit below 10%

(B) moisture content below 10%

(C) organic content below 10%

(D) plastic index above 40%

210. One laboratory test used to indicate the grain size of fine-grained soil is the

(A) sieve analysis

(B) Atterberg limit test

(C) hydrometer analysis

(D) California bearing ratio (CBR) test

SITUATION FOR PROBLEMS 211–215

A field inspection of an asphalt paving site determines that the mix temperature is 270°F. The first layer thickness is 3 in. The paving train contains pneumatic and vibratory rollers primarily. A nuclear density gage determines that the dry density is 140.2 lbf/ft^3, and the bulk density is 145.2 lbf/ft^3. A Marshall test determines that the maximum specific gravity of the mixture is 2.5.

LA 0000

211. What is the approximate relative compaction?

(A) 91%

(B) 93%

(C) 95%

(D) 97%

212. The paving contractor has complained about the initial layer thickness. For achieving proper compaction, what should be the maximum lift of the initial layer?

(A) 3 in

(B) 5 in

(C) 9 in

(D) 12 in

213. The three factors influencing the compaction of asphalt pavement mixtures most are

  I. ambient temperature
  II. asphalt grade
  III. lift thickness
  IV. mixture temperature
  V. number of roller passes
  VI. percent crown
  VII. percent RAP in the mixture
  VIII. speed of paver

(A) III, IV, and V

(B) II, III, and VII

(C) V, VI, and VIII

(D) I, II, and VII

214. The percentage asphalt cement content in asphalt concrete pavement mixtures for highway use is usually

(A) between 2% and 4%

(B) between 4% and 6%

(C) greater than 10%

(D) less than 10%

215. The nuclear gauge's percent moisture dial reads 5.2% when placed on top of the hot, new pavement. This represents the percentage of

(A) water in the pavement

(B) water-based emulsifier in the pavement

(C) moisture in the air

(D) hydrocarbons in the mixture

SITUATION FOR PROBLEMS 216–220

2.5 in high by 4 in diameter asphalt concrete "biscuits" are prepared using the 50-blow Marshall method. The maximum specific gravity of the asphalt concrete is 2.56. After one particular biscuit has cooled, the following measurements are obtained from it.

air dry mass	1210 g	A
submerged mass	715 g	C
SSD mass	1215 g	B

$G_{mm} = 2.56$ *LA 0000*

216. For the purposes of the Marshall method, the bulk volume of the asphalt is

(A) 495 cm^3

(B) 500 cm^3

(C) 1210 cm^3

(D) 1215 cm^3

217. The approximately bulk specific gravity of the asphalt is

(A) 2.38

(B) 2.40

(C) 2.42

(D) 2.43

218. The approximate bulk unit weight of the asphalt is

(A) 149 lbf/ft^3

(B) 151 lbf/ft^3

(C) 153 lbf/ft^3

(D) 155 lbf/ft^3

219. The relative compaction of the asphalt is

(A) 94.5%

(B) 95.5%

(C) 96.5%

(D) 97.5%

220. The percent air voids in the asphalt is approximately

(A) 3.5%

(B) 4.5%

(C) 5.5%

(D) 6.5%

SITUATION FOR PROBLEMS 221–230

A concrete retaining wall with a 23 ft overall height supports a noncohesive backfill. The footing base does not have a key. The backfill has an average unit weight of 130 lbf/ft^3 and an angle of internal friction of 35°. The soil is flush with the top of the wall and horizontal behind it. The coefficient of sliding friction between the soil and bottom of the footing base is 0.60. Passive pressure distributions are negligible.

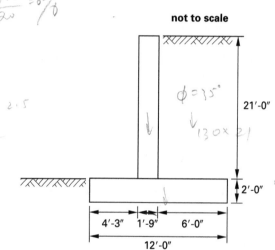

221. According to the Rankine theory, what is the coefficient of active earth pressure?

(A) 0.18

(B) 0.21

(C) 0.27

(D) 0.32

222. What is the approximate resultant active force acting on the stem per foot of wall width?

(A) 7.8 kips/ft

(B) 9.2 kips/ft

(C) 12 kips/ft

(D) 16 kips/ft

223. What is the approximate active bending moment per foot of wall width at the base of the stem?

(A) 54 ft-kips/ft

(B) 62 ft-kips/ft

(C) 71 ft-kips/ft

(D) 89 ft-kips/ft

224. What is the approximate shear force acting per foot of retaining wall width at the base of the stem?

(A) 6.3 kips/ft

(B) 7.8 kips/ft

(C) 8.5 kips/ft

(D) 9.2 kips/ft

225. What is the approximate overturning moment per foot of wall width?

(A) 43 ft-kips/ft

(B) 54 ft-kips/ft

(C) 60 ft-kips/ft

(D) 71 ft-kips/ft

226. What is the approximate stabilizing moment per foot of wall width?

(A) 71 ft-kips/ft

(B) 98 ft-kips/ft

(C) 140 ft-kips/ft

(D) 200 ft-kips/ft

227. What is the approximate factor of safety against overturning?

(A) 1.5

(B) 2.2

(C) 2.8

(D) 3.1

228. What is the approximate margin of safety against sliding?

(A) 0.6

(B) 1.2

(C) 1.6

(D) 1.9

229. What is the approximate maximum soil pressure acting on the retaining wall base?

(A) 1.1 kips/ft^2

(B) 2.3 kips/ft^2

(C) 2.7 kips/ft^2

(D) 3.3 kips/ft^2

230. What is the approximate minimum soil pressure acting on the retaining wall base?

(A) 0.5 kips/ft^2

(B) 0.8 kips/ft^2

(C) 1.0 kips/ft^2

(D) 1.3 kips/ft^2

SITUATION FOR PROBLEMS 231–240

A 3,000,000 yd^3 earthen dam is constructed from borrow soil with the following original properties.

void ratio	0.70
degree of saturation	0.40
solids specific gravity	2.65

The borrow soil is compacted to a final specific weight of 110 lbf/ft^3 and a 17% moisture content. The trucks used to transport the borrow soil from the borrow pit to the dam site each have maximum capacities of 10 yd^3 and 14 tons.

231. What is the approximate weight of solids in 1 ft^3 of compacted fill?

(A) 92 pounds or less

(B) 93 pounds

(C) 94 pounds

(D) 95 pounds or more

232. What is the approximate weight of water in 1 ft^3 of compacted fill?

(A) 15 pounds or less

(B) 16 pounds

(C) 17 pounds

(D) 18 pounds or more

233. What is the approximate volume of voids in 1 ft^3 of undisturbed borrow soil?

(A) 0.42 ft^3 or less

(B) 0.45 ft^3

(C) 0.51 ft^3

(D) 0.55 ft^3 or more

234. What is the approximate volume of solids in 1 ft^3 of undisturbed borrow soil?

(A) 0.41 ft^3 or less

(B) 0.47 ft^3

(C) 0.50 ft^3

(D) 0.58 ft^3 or more

235. What is the approximate volume of water in 1 ft^3 of undisturbed borrow soil?

(A) 0.10 ft^3 or less

(B) 0.12 ft^3

(C) 0.14 ft^3

(D) 0.16 ft^3 or more

236. What is the approximate weight of solids in 1 ft^3 of undisturbed borrow soil?

(A) 89 pounds or less

(B) 94 pounds

(C) 97 pounds

(D) 99 pounds or more

237. What is the approximate weight of water in 1 ft^3 of undisturbed borrow soil?

(A) 8 pounds or less

(B) 9 pounds

(C) 10 pounds

(D) 11 pounds or more

238. Approximately what volume of borrow soil is required?

(A) 2.7×10^6 yd^3 or less

(B) 2.9×10^6 yd^3

(C) 3.1×10^6 yd^3

(D) 3.3×10^6 yd^3 or more

239. When loading into the trucks, approximately how much will the volume of the borrow soil expand?

(A) 1% or less

(B) 5%

(C) 15%

(D) 25% or more

240. Given a fluff factor of 10%, approximately how many trips must the truck make to complete the dam?

(A) 290,000 trips or fewer

(B) 300,000 trips

(C) 310,000 trips

(D) 320,000 trips or more

STOP!

DO NOT CONTINUE!

This concludes the Afternoon Session of the examination. If you finish early, check your work and make sure that you have followed all instructions. After checking your answers, you may turn in your Examination Booklet and Answer Sheet and leave the examination room. Once you leave, you will not be permitted to return to work or change your answers.

Solutions
Morning Session

1. The watershed slope for subbasin 1 is

$$L = (2.12 \text{ mi} + 0.50 \text{ mi}) \left(5280 \frac{\text{ft}}{\text{mi}} \right)$$

$$= 13{,}834 \text{ ft}$$

$$S_0 = \frac{\Delta z}{L} = \frac{1130 \text{ ft} - 864 \text{ ft}}{13{,}834 \text{ ft}}$$

$$= 0.01923 \quad (1.923\%)$$

The answer is C.

2. Since the natural condition of the basin in desert, HSG A (high infiltration rates) is appropriate.

The answer is A.

3. For ARC II, HSG A, and residential districts with $\frac{1}{8}$ acre townhouses, CN = 77.

The answer is B.

4. The NRCS (SCS) method calculates the storage capacity from the curve number. (The symbol for storage capacity is the same as for slope. However, the meanings are different.)

$$S = \frac{1000}{\text{CN}} - 10 = \frac{1000}{77} - 10$$

$$= 2.987 \text{ in}$$

The answer is C.

5. The NRCS (SCS) method calculates the time lag as

$$t_L = \frac{L^{0.8}(S+1)^{0.7}}{1900\sqrt{S_0}}$$

$$= \frac{(13{,}834 \text{ ft})^{0.8}(2.987+1)^{0.7}}{1900\sqrt{1.923\%}}$$

$$= 2.05 \text{ hr}$$

The answer is D.

6. From the AISC Manual specification Sec. I1, the effective beam width is

$$b = \text{smaller of} \left\{ \frac{L}{4}, \ s \right\}$$

$$= \text{smaller of} \left\{ \frac{25 \text{ ft}}{4} = 6.25 \text{ ft}, \ 5 \text{ ft} \right\}$$

$$= 5 \text{ ft}$$

The answer is C.

7. One of the Whitney assumptions is that the steel is stressed in tension to its yield point at ultimate capacity. The tensile area of an W8×48 beam is 14.1 in^2.

$$F_T = f_y A_s = \left(36{,}000 \ \frac{\text{lbf}}{\text{in}^2} \right) (14.1 \text{ in}^2)$$

$$= 507{,}600 \text{ lbf}$$

The answer is D.

8. From the wording of the question, the neutral axis is within the concrete. (This is a common first assumption, in any case.) None of the steel is in compression. The top 1 in of the concrete is in compression, where a = height of the stress block. The concrete is stressed to 85% of its compressive strength. The resultant compressive force in the concrete is

$$F_C = 0.85 f_c' A_c = 0.85 f_c' a b$$

For equilibrium, the compressive force is equal to the tensile force. Therefore,

$$F_C = F_T = 507{,}600 \text{ lbf}$$

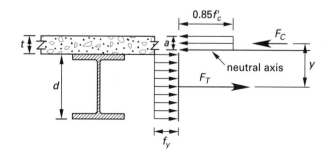

The height of the compressive block is

$$a = \frac{F_C}{0.85 f'_c b} = \frac{F_T}{0.85 f'_c b}$$

$$= \frac{507{,}600 \text{ lbf}}{(0.85)\left(4000 \dfrac{\text{lbf}}{\text{in}^2}\right)(5 \text{ ft})\left(12 \dfrac{\text{in}}{\text{ft}}\right)}$$

$$= 2.488 \text{ in}$$

The answer is A.

9. The depth, d, of a $W 8 \times 48$ beam is 8.5 in.

The distance between the tensile and compressive resultants is

$$y = \frac{d}{2} + t - \frac{a}{2}$$

$$= \frac{8.5 \text{ in}}{2} + 4 \text{ in} - \frac{2.488 \text{ in}}{2}$$

$$= 7.006 \text{ in}$$

The answer is B.

10. The ultimate moment capacity is

$$M_u = \phi F_C y = \phi F_T y$$

$$= \frac{(0.90)(507{,}600 \text{ lbf})(7.006 \text{ in})}{12 \dfrac{\text{in}}{\text{ft}}}$$

$$= 2.667 \times 10^5 \text{ ft-lbf}$$

The moment from a distributed load is

$$M = \frac{wL^2}{8}$$

The ultimate uniform load is

$$w_u = \frac{8 M_u}{L^2} = \frac{(8)(2.667 \times 10^5 \text{ ft-lbf})}{(25 \text{ ft})^2}$$

$$= 3414 \text{ lbf/ft}$$

The actual dead load per foot of girder is

$$w_d = 48 \frac{\text{lbf}}{\text{ft}} + \left(\frac{(4 \text{ in})(5 \text{ ft})\left(150 \dfrac{\text{lbf}}{\text{ft}^3}\right)}{12 \dfrac{\text{in}}{\text{ft}}} \right)$$

$$= 298 \text{ lbf/ft}$$

Using 1.4 and 1.7 as the load factors for dead and live loads, respectively, the maximum live load is

$$w_l = \frac{w_u - 1.4 w_d}{1.7}$$

$$= \frac{3414 \dfrac{\text{lbf}}{\text{ft}} - (1.4)\left(298 \dfrac{\text{lbf}}{\text{ft}}\right)}{1.7}$$

$$= 1763 \text{ lbf/ft}$$

(The *Manual of Steel Construction* (LRFD Design) uses values of 1.2 and 1.6 for dead and live loads, respectively.)

The answer is C.

11. The weight of the footing must be added to the dead load. Since the footing thickness is initially unknown, a 24 in thickness is used.

The allowable soil pressure does not require a factor of safety. The net allowable soil pressure is

$$q_{\text{net}} = p_a - t \gamma_{\text{concrete}}$$

$$= 3500 \frac{\text{lbf}}{\text{ft}^2} - \frac{(24 \text{ in})\left(150 \dfrac{\text{lbf}}{\text{ft}^3}\right)}{12 \dfrac{\text{in}}{\text{ft}}}$$

$$= 3200 \text{ lbf/ft}^2$$

The applied load is not factored in calculating the footing size. [ACI 318 Sec. 15.2.2] The approximate area required is

$$A = \frac{P_d + P_l}{q_{\text{net}}}$$

$$= \frac{125{,}000 \text{ lbf} + 175{,}000 \text{ lbf}}{3200 \dfrac{\text{lbf}}{\text{ft}^2}}$$

$$= 93.75 \text{ ft}^2$$

Since the footing is square, try a footing with sides of 10 ft.

The answer is B.

12. The thickness allows for two layers of tension steel (one layer in each direction, as well as 3 in of cover on the bottom steel). The tension bars will be approximately 1 inch in diameter. The distance from the extreme compression fiber (the top surface of the footing) to the centerline of the top layer of tension reinforcement is

$$d = t - d_b - \frac{d_b}{2} - \text{cover}$$

$$= 24 \text{ in} - 1 \text{ in} - \frac{1 \text{ in}}{2} - 3 \text{ in}$$

$$= 19.5 \text{ in} \quad \left[\begin{array}{c} \text{Use } d = 19 \text{ in for} \\ \text{preliminary estimates.} \end{array} \right]$$

The critical section for two-way punching shear starts at a distance $d/2$ from the column face.

$$\frac{d}{2} = \frac{19 \text{ in}}{2} = 9.5 \text{ in}$$

The length of each side of the inner periphery of the two-way punching shear area is

$$l_o = 9.5 \text{ in} + 14 \text{ in} + 9.5 \text{ in} = 33 \text{ in}$$

The critical perimeter length is

$$b_o = 4l_o = (4)(33 \text{ in}) = 132 \text{ in}$$

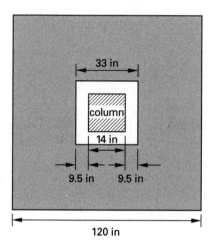

The critical area for shear force lies outside the critical perimeter. The critical area is

$$A_{\text{critical}} = b_w^2 - l_o^2 = \frac{(120 \text{ in})^2 - (33 \text{ in})^2}{\left(12 \; \frac{\text{in}}{\text{ft}}\right)^2}$$

$$= 92.44 \text{ ft}^2$$

The answer is D.

13. The nominal concrete shear strength is given by ACI 318 Sec. 11.12.2.1.

$$\beta_c = \frac{b_c}{a_c} = \frac{14 \text{ in}}{14 \text{ in}} = 1$$

$$\beta_c \geq 2$$

$$V_c = (2 + y)\sqrt{f'_c}b_o d$$

$$y = \min\left\{2, \; \frac{4}{\beta_c}, \; \frac{40d}{b_o}\right\}$$

$$= \min\left\{2, \; \frac{4}{2} = 2, \; \frac{(40)(19 \text{ in})}{132 \text{ in}} = 5.8\right\}$$

$$= 2$$

(Even though the footing is located at a corner of the building, it is concentric with the column and has four critical sides. Therefore, it is an "interior" column from the stand point of ACI 318 Sec. 11.12.2.2.)

$$V_c = (2 + 2)\sqrt{f'_c}b_o d$$

$$= (4)\sqrt{3000 \; \frac{\text{lbf}}{\text{in}^2}}(132 \text{ in})(19 \text{ in})$$

$$= 549{,}475 \text{ lbf}$$

The answer is C.

14. The footing area is

$$A = B^2 = (10 \text{ ft})(10 \text{ ft}) = 100 \text{ ft}^2$$

The ultimate load carried by the footing is

$$P_u = 1.4P_d + 1.7P_l$$

$$= (1.4)(125{,}000 \text{ lbf}) + (1.7)(175{,}000 \text{ lbf})$$

$$= 472{,}500 \text{ lbf}$$

The factored soil pressure acting on the footing is

$$p_u = \frac{P_u}{A} = \frac{472{,}500 \text{ lbf}}{100 \text{ ft}^2}$$

$$= 4725 \text{ lbf/ft}^2$$

The ultimate two-way punching shear is

$$V_u = p_u A_{\text{critical}} = \left(4725 \; \frac{\text{lbf}}{\text{ft}^2}\right)(92.44 \text{ ft}^2)$$

$$= 436{,}779 \text{ lbf}$$

The answer is B.

15. The critical area for beam action is located a distance d from the column face.

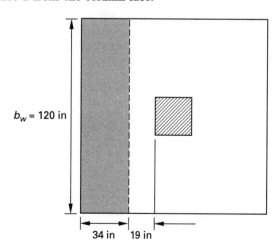

The length of the critical section is

$$l = b_w - \frac{b_w}{2} - \frac{b_c}{2} - d$$

$$= 120 \text{ in} - \frac{120 \text{ in}}{2} - \frac{14 \text{ in}}{2} - 19 \text{ in}$$

$$= 34 \text{ in}$$

The critical area for beam action shear force is

$$A_{\text{critical}} = b_w l = \frac{(120 \text{ in})(34 \text{ in})}{\left(12 \dfrac{\text{in}}{\text{ft}}\right)^2}$$

$$= 28.3 \text{ ft}^2$$

The answer is B.

16. The traffic is predominantly from A to B and B to A. The turning movements can be accommodated by at-grade turns. A simple diamond is recommended. (Note that ADT can be converted to useful hourly traffic volumes by multiplying by a factor of approximately 0.15.)

The answer is C.

17. The traffic is heavy to and from all direct destinations. All of the turning movements are also fairly high. Ideally, the hourly turning movements are within the high-end range of diamond intersections. However, there is little room for expansion of the average traffic, and morning/evening commutes undoubtedly will exceed the ideal capacity. A full cloverleaf is recommended.

The answer is A.

18. The traffic is very heavy to and from all direct destinations except D. The turning movements are too numerous to be accommodated by a cloverleaf, which would require significant weaving. Some type (e.g., three-leg) of directional interchange is recommended. Loop ramps can be used to accommodate the AD/DA flow.

The answer is D.

19. 1.0 in of runoff from the watershed will produce a volume of

$$V = A_d h = \frac{(43 \text{ mi}^2)\left(640 \dfrac{\text{ac}}{\text{mi}^2}\right)(1 \text{ in})}{12 \dfrac{\text{in}}{\text{ft}}}$$

$$= 2293 \text{ ac-ft}$$

Since the actual flood hydrograph volume was 3260 ac-ft, the depth of runoff in the flood hydrograph was

$$d = \frac{V_{\text{actual}}}{V_{\text{unit}}} = \frac{3260 \text{ ac-ft}}{2293 \dfrac{\text{ac-ft}}{\text{in}}}$$

$$= 1.42 \text{ in}$$

The actual flood hydrograph peak was 9300 ft^3/sec, so the unit hydrograph peak is

$$Q_{p,\text{unit}} = \frac{Q_{p,\text{actual}}}{d} = \frac{9300 \dfrac{\text{ft}^3}{\text{sec}}}{1.42 \text{ in}}$$

$$= 6550 \text{ ft}^3/\text{sec-in} \quad \text{[for a 1 in storm]}$$

The answer is D.

20. The design flood hydrograph volume from a 2.5 in storm is

$$\left(2293 \frac{\text{ac-ft}}{\text{in}}\right)(2.5 \text{ in}) = 5733 \text{ ac-ft}$$

The answer is B.

21. Total inorganic carbon (TIC) is calculated from the concentrations as

$$\text{TIC} = [\text{HCO}_3^-] + [\text{CO}_3^{--}]$$

CO_2 and Na_2CO_3 both contribute CO_3^{--}. The other compounds do not contain carbon.

The answer is C.

22. Increasing the pH is the same as becoming more basic. NaOH is a base. CO_2 makes carbonic acid, lowering the pH. Na_2CO_3 does not change the pH. HCl is an acid, which lowers the pH.

The answer is A.

23. Given the compounds being considered, alkalinity is calculated from the concentrations as

$$M = [\text{OH}^-] + 2[\text{CO}_3^{--}] + [\text{HCO}_3^-] - [\text{H}^+]$$

All compounds contribute one of the above ions.

The answer is D.

24. If added to the water, Na_2CO_3 will increase alkalinity and total inorganic carbon. There will be no effect on pH.

The answer is D.

25. If added to the water, HCl will decrease alkalinity and pH. There will be no effect on inorganic carbon, because there is no carbon in HCl.

The answer is B.

26. The fraction of suspended solids removed is

$$\eta = \frac{C_i - C_o}{C_i} = \frac{500\ \frac{mg}{L} - 150\ \frac{mg}{L}}{500\ \frac{mg}{L}}$$

$$= 0.70$$

The answer is C.

27. 10 MGD is the average flow/capacity. The primary clarifier system must have a capacity greater than 10 MGD to perform during times of peak flow. In reality, the peaking factor depends on the nature of the source. In this problem, the peaking factor used will depend on the authority cited. Values of 1.5 to 2.0 are reasonable.

$$Q = \frac{(1.5)(10\ \text{MGD})\left(1,000,000\ \frac{gal}{MG}\right)}{24\ \frac{hr}{day}}$$

$$= 625,000\ \text{gal/hr}$$

The answer is C.

28. Primary clarifiers should be designed primarily for overflow rate. From the pilot study data, for a suspended solids removal of 70%, the overflow rate is approximately 1900 gpd/ft^2.

$$v^* = \frac{Q}{A} = \frac{Q}{\left(\frac{\pi}{4}\right)D^2}$$

$$D = \sqrt{\frac{4Q}{\pi v^*}}$$

$$= \sqrt{\frac{(4)(15\ \text{MGD})\left(1,000,000\ \frac{gal}{MG}\right)}{\pi\left(1900\ \frac{gal}{day\text{-}ft^2}\right)}}$$

$$= 100.3\ \text{ft} \quad [\text{use } 100\ \text{ft}]$$

The answer is D.

29. From the pilot study, the detention time for a 70% suspended solids removal is approximately 33 min. The tank volume is

$$V = t_d Q$$

$$= \frac{(33\ \text{min})(15\ \text{MGD})\left(1,000,000\ \frac{gal}{MG}\right)}{\left(24\ \frac{hr}{day}\right)\left(60\ \frac{min}{hr}\right)(2\ \text{units})\left(7.48\ \frac{gal}{ft^3}\right)}$$

$$= 22{,}978\ \text{ft}^3/\text{unit}$$

The depth in each unit is

$$d = \frac{V}{A} = \frac{V}{\left(\frac{\pi}{4}\right)D^2}$$

$$= \frac{22{,}978\ \text{ft}^3}{\left(\frac{\pi}{4}\right)(72\ \text{ft})^2}$$

$$= 5.64\ \text{ft}$$

The answer is B.

30. The weir loading rate is

$$\text{weir loading} = \frac{Q}{L} = \frac{Q}{\pi D}$$

$$= \frac{(15\text{MGD})\left(1,000,000\ \frac{gal}{MG}\right)}{(2\ \text{units})\pi(72\ \text{ft})}$$

$$= 33{,}157\ \text{gal/day-ft}$$

The answer is A.

31. The test hole log is labeled "Hole #3." Therefore, there were at least 3 holes dug.

The answer is C.

32. "SS" indicates that a split-spoon was used to collect the sample.

The answer is A.

33. The symbol for the soil encountered at a depth of 1 m, diagonal lines with specs and a vertical line, indicates the soil was till (sandy clay). All of the other choices have distinctly different symbols.

The answer is D.

34. The penetration test at a depth of 2.5 m shows that the N-value is approximately 20. This falls in the middle of the "medium dense" category. (Note that the N-value for "dense" is greater than that for "medium dense.")

The answer is C.

35. The log indicates that a field vane test was performed at a depth of approximately 1.6 m. The shear strength (cohesion) was found to be 30 kPa. For saturated clay, the undrained shear strength, S_u, is the same as the cohesion.

The answer is B.

36. The actual load on the soil is

$$r = \frac{D}{2} = \frac{40 \text{ ft}}{2}$$
$$= 20 \text{ ft}$$

$$p_{actual} = \frac{P}{A} = \frac{P}{\pi r^2} = \frac{2{,}500{,}000 \text{ lbf}}{\pi (20 \text{ ft})^2}$$
$$= 1989 \text{ lbf/ft}^2$$

The factor of safety is

$$F = \frac{q_a}{p_{actual}} = \frac{3000 \dfrac{\text{lbf}}{\text{ft}^2}}{1989 \dfrac{\text{lbf}}{\text{ft}^2}}$$
$$= 1.51$$

The answer is B.

37. The vertical stress at the midpoint of the clay layer is the sum of the effective stresses at that point.

$$p_o = \sum \gamma' H$$
$$= \left(130 \frac{\text{lbf}}{\text{ft}^3} \right) (5 \text{ ft}) + \left(132.4 \frac{\text{lbf}}{\text{ft}^3} - 62.4 \frac{\text{lbf}}{\text{ft}^3} \right) (5 \text{ ft})$$
$$= 1000 \text{ lbf/ft}^2$$

The answer is B.

38. The increase in vertical stress can be found in a number of ways. Each method will produce a slightly different answer. However, the answer choices are significantly different, so the method used probably will not matter. In this case, it is noted that the depth (10 ft) to the point of interest is much less than the diameter (40 ft), so the increase in vertical pressure will be close to the applied pressure.

$$\Delta p_v = 1989 \frac{\text{lbf}}{\text{ft}^2}$$
$$p_v = p_o + \Delta p_v$$
$$= 1000 \frac{\text{lbf}}{\text{ft}^2} + 1989 \frac{\text{lbf}}{\text{ft}^2}$$
$$= 2989 \text{ lbf/ft}^2$$

The answer is C.

39. The total clay thickness is 10 ft. The primary settlement is

$$S = \frac{H C_c \log_{10} \left(\dfrac{p_o + \Delta p_v}{p_o} \right)}{1 + e_o}$$
$$= \frac{(10 \text{ ft})(0.34) \log_{10} \left(\dfrac{1000 \dfrac{\text{lbf}}{\text{ft}^2} + 1989 \dfrac{\text{lbf}}{\text{ft}^2}}{1000 \dfrac{\text{lbf}}{\text{ft}^2}} \right)}{1 + 1.15}$$
$$= 0.752 \text{ ft}$$

The answer is D.

40. The time factor for 80% settlement is approximately 0.567. The required time is

$$t = \frac{T_v H^2}{C_v} = \frac{(0.567)(10 \text{ ft})^2}{0.10 \dfrac{\text{ft}^2}{\text{day}}}$$
$$= 567 \text{ days}$$

The answer is D.

Solutions
Water Resources

41. The total length of the stream is

$$L = (21 \text{ sta}) \left(100 \; \frac{\text{ft}}{\text{sta}} \right) + 35.69 \text{ ft}$$
$$= 2135.69 \text{ ft}$$

The geometric slope is

$$S_0 = \frac{\Delta z}{L} = \frac{1.54 \text{ ft}}{2135.69 \text{ ft}}$$
$$= 0.00072$$

The answer is A.

42. Although the value can vary somewhat, a value of $n = 0.035$ is consistent with a natural intermittent stream.

The answer is C.

43. The maximum flow area is

$$A = wd = (4 \text{ ft})(6 \text{ ft})$$
$$= 24 \text{ ft}^2$$

The wetted perimeter is

$$P = w + 2d = 4 \text{ ft} + (2)(6 \text{ ft})$$
$$= 16 \text{ ft}$$

The hydraulic radius is

$$R = \frac{A}{P} = \frac{24 \text{ ft}}{16 \text{ ft}}$$
$$= 1.5 \text{ ft}$$

Use the U.S. form of the Chezy-Manning equation.

$$Q = \left(\frac{1.49}{n} \right) AR^{\frac{2}{3}} \sqrt{S_0}$$
$$= \left(\frac{1.49}{0.035} \right) (24 \text{ ft}^2)(1.5 \text{ ft})^{\frac{2}{3}} \sqrt{0.00072}$$
$$= 35.9 \text{ ft}^3/\text{sec}$$

The answer is A.

44. The answer will depend on the reference used. From the *Civil Engineering Reference Manual* (Lindeburg) for erodible "firm earth, small channels," a side slope ratio of 1.5:1 [horizontal:vertical] is recommended.

The answer is B.

45. *Method 1: Algebraic*

The Chezy-Manning equation relates the flow rate to the hydraulic properties of the channel.

$$Q = \left(\frac{1.49}{n} \right) AR^{\frac{2}{3}} \sqrt{S_0} = \frac{K' b^{\frac{8}{3}} \sqrt{S_0}}{n}$$

The flow area, A, and the hydraulic radius, R, are both related to the depth of flow. Use a solution based on the conveyance, K'.

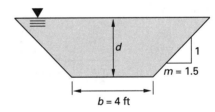

$$K' = \frac{(1.49) \left[1 + m \left(\dfrac{d}{b} \right) \right]^{\frac{5}{3}} \left(\dfrac{d}{b} \right)^{\frac{5}{3}}}{\left(1 + (2) \left(\dfrac{d}{b} \right) \sqrt{1 + m^2} \right)^{\frac{2}{3}}}$$
$$m = 1.5$$

Choose $d/b = 0.7$ (roughly based on a flow depth of 3 ft in a 4 ft wide channel). Then,

$$K' = \frac{(1.49) \left(1 + (1.5)(0.7) \right)^{\frac{5}{3}} (0.7)^{\frac{5}{3}}}{\left(1 + (2)(0.7) \sqrt{1 + (1.5)^2} \right)^{\frac{2}{3}}}$$
$$= (1.49)(0.788)$$
$$= 1.17$$

Choose $d/b = 1.2$ (roughly based on a flow depth of 5 ft in a 4 ft wide channel). Then,

$$K' = \frac{(1.49)\left(1 + (1.5)(1.2)\right)^{\frac{5}{3}} (1.2)^{\frac{5}{3}}}{\left(1 + (2)(1.2)\sqrt{1 + (1.5)^2}\right)^{\frac{2}{3}}}$$
$$= (1.49)(2.47)$$
$$= 3.68$$

The actual value of K' is

$$K' = \frac{nQ}{b^{\frac{8}{3}}\sqrt{S_0}} = \frac{(0.018)\left(138 \, \frac{\text{ft}^3}{\text{sec}}\right)}{(4 \text{ ft})^{\frac{8}{3}}\sqrt{0.00072}}$$
$$= 2.30$$

Using straight-line interpolation,

$$\frac{3.68 - 1.17}{1.2 - 0.7} = \frac{2.30 - 1.17}{\dfrac{d}{b} - 0.7}$$
$$\frac{d}{b} = 0.93$$
$$d = 0.93b = (0.93)(4 \text{ ft})$$
$$= 3.72 \text{ ft}$$

(The interpolation is not truly straight-line. A better estimate of d/b is 0.97.)

Method 2: Using tabulated values or graphs

The actual value of K' is

$$K' = \frac{nQ}{b^{\frac{8}{3}}\sqrt{S_0}} = \frac{(0.018)\left(138 \, \frac{\text{ft}^3}{\text{sec}}\right)}{(4 \text{ ft})^{\frac{8}{3}}\sqrt{0.00072}}$$
$$= 2.30$$

(Note that in metric tables, $K' = 2.30/1.49 = 1.54$ would be the target K' value.)

Using trapezoidal channel tables of K' versus d/b and m in *Handbook of Hydraulics* (Brater, King, Lindell, and Wei), with $K' = 2.30$, $m = 1\frac{1}{2}{:}1$, $d/b = 0.97$.

$$d = 0.97b = (0.97)(4 \text{ ft})$$
$$= 3.72 \text{ ft}$$

The answer is B.

46. When the depth of flow is 3.9 ft, the area in flow will be

$$A = db + (2)\tfrac{1}{2}md^2$$
$$= (3.9 \text{ ft})(4 \text{ ft}) + (2)\left(\tfrac{1}{2}\right)(1.5)(3.9 \text{ ft})^2$$
$$= 38.42 \text{ ft}^2$$

The wetted perimeter is

$$P = b + 2\sqrt{d^2 + (1.5d)^2}$$
$$= 4 \text{ ft} + 2\sqrt{(3.9 \text{ ft})^2 + \left((1.5)(3.9 \text{ ft})\right)^2}$$
$$= 18.06 \text{ ft}$$

The hydraulic radius is

$$R = \frac{A}{P} = \frac{38.42 \text{ ft}^2}{18.06 \text{ ft}}$$
$$= 2.13 \text{ ft}$$

Use the Chezy-Manning equation.

$$v = \left(\frac{1.49}{n}\right) R^{\frac{2}{3}}\sqrt{S_0}$$
$$= \left(\frac{1.49}{0.018}\right)(2.13 \text{ ft})^{\frac{2}{3}}\sqrt{0.00072}$$
$$= 3.68 \text{ ft/sec}$$

The answer is C.

47. When the depth of flow is 6 ft, the area in flow will be

$$A = db + (2)\left(\tfrac{1}{2}md^2\right)$$
$$= (6 \text{ ft})(4 \text{ ft}) + (2)\left(\tfrac{1}{2}\right)(1.5)(6 \text{ ft})^2$$
$$= 78.0 \text{ ft}^2$$

The wetted perimeter is

$$P = b + 2\sqrt{d^2 + (1.5d)^2}$$
$$= 4 \text{ ft} + 2\sqrt{(6 \text{ ft})^2 + \left((1.5)(6 \text{ ft})\right)^2}$$
$$= 25.63 \text{ ft}$$

The hydraulic radius is

$$R = \frac{A}{P} = \frac{78.0 \text{ ft}^2}{25.63 \text{ ft}}$$
$$= 3.04 \text{ ft}$$

Use the Chezy-Manning equation.

$$Q = vA = \left(\frac{1.49}{n}\right) AR^{\frac{2}{3}}\sqrt{S_0}$$
$$= \left(\frac{1.49}{0.018}\right)(78.0 \text{ ft}^2)(3.04 \text{ ft})^{\frac{2}{3}}\sqrt{0.00072}$$
$$= 364 \text{ ft}^3/\text{sec}$$

The answer is B.

48. *Method 1: Analytic (trial and error)*

For critical flow in a trapezoidal channel with a surface width of b_s,

$$\frac{A^3}{b_s} = \frac{Q^2}{g} = \frac{\left(138\ \frac{\text{ft}^3}{\text{sec}}\right)^2}{32.2\ \frac{\text{ft}}{\text{sec}^2}}$$

$$= 591\ \text{ft}^5$$

$$A = db + (2)\left(\tfrac{1}{2}md^2\right)$$
$$= d(4\ \text{ft}) + (2)\left(\tfrac{1}{2}\right)(1.5)d^2$$
$$= 4d + 1.5d^2$$

$$b_s = b + 2md$$
$$= 4\ \text{ft} + (2)(1.5)d$$
$$= 4 + 3d$$

Use trial and error.

d	A^3/b_s
3.2	1640 (> 591)
1.6	122 (< 591)
2.4	542 (< 591)
2.6	734 (> 591)

The critical depth is approximately 2.45 ft.

Method 2: Using tabulated values or graphs

The critical conveyance is

$$K_c' = \frac{Q}{b^{\frac{5}{2}}} = \frac{138\ \frac{\text{ft}^3}{\text{sec}}}{(4\ \text{ft})^{\frac{5}{2}}}$$

$$= 4.31$$

(Note that in metric tables, $K_c' = 4.31/(1.49)^{\frac{3}{2}} = 2.37$ would be the target K_c' value.)

Using trapezoidal channel tables of K_c' versus d/b and m in *Handbook of Hydraulics* (Brater, King, Lindell, and Wei), with $K_c' = 4.31$, $d/b = 0.615$, and $m = 1\frac{1}{2}{:}1$,

$$d = 0.615b = (0.615)(4\ \text{ft})$$
$$= 2.46\ \text{ft}$$

The answer is B.

49. The Chezy-Manning equation shows that the flow rate is inversely proportional to the roughness coefficient, n. Since $n_{\text{riprap}} > n_{\text{earth}}$, the capacity will decrease. In order to carry the same flow rate, the depth must increase.

The answer is B.

50. Manning's roughness coefficient for canals whose sides and bottom are lined with irregularly cut rock is approximately 0.035.

The answer is C.

51. Refer to a table of fitting losses. Answers will vary depending on the reference used, since each manufacturer's fittings are somewhat different. This should not affect the answer choice too much, since the selections differ from one another by approximately 100 ft. Values used in this solution are taken from the *Civil Engineering Reference Manual* (Lindeburg).

pipe run		2825 ft
elbows	(4)(5.7 ft) =	22.8 ft
swing check valves	(2)(63 ft) =	126 ft
gate valve	(1)(3.2 ft) =	3.2 ft

Data on butterfly valves is more elusive. The loss coefficient for butterfly valves in the 2 in to 8 in range is approximately $45f_t$, where f_t is the pipe friction factor for fully turbulent flow. For 6 in schedule-40 steel pipe,

$$A_i = 0.2006\ \text{ft}^2$$
$$\epsilon = 0.0002\ \text{ft}$$
$$D_i = 0.5054\ \text{ft}$$

The relative roughness is

$$\frac{\epsilon}{D_i} = \frac{0.0002\ \text{ft}}{0.5054\ \text{ft}} \approx 0.0004$$

From a Moody friction factor chart for a relative roughness of $\epsilon/D = 0.0004$, $f = 0.016$ in the fully turbulent region.

The loss coefficient for the butterfly valve is

$$K = 45f_t = (45)(0.016)$$
$$= 0.72$$

The actual friction factor, f, is not yet known. (It is not the same as the friction factor for fully turbulent flow.) A reasonable assumption is required for the first iteration. In this case, the fully turbulent value ($f = 0.016$) is used. When the operating point is determined, it will be possible to check the actual friction factor. However, it is unlikely that this calculation is going to affect the operating point significantly.

The relationship between the loss coefficient and the equivalent length is

$$L_e = \frac{KD_i}{f} = \frac{(0.72)(0.5054\ \text{ft})}{0.016}$$
$$= 22.7\ \text{ft}$$

Venturi meters have gradual inlets and outlets. Their friction losses are very low. Therefore, no additional length is included for the venturi meter.

The approximate total equivalent length is

$$L_e = 2825 \text{ ft} + 22.8 \text{ ft} + 126 \text{ ft} + 3.2 \text{ ft} + 22.7 \text{ ft}$$
$$= 2999.7 \text{ ft} \quad [\text{use 3000 ft}]$$

The answer is C.

52. Construct the pump characteristic curve from the pump manufacturer's data.

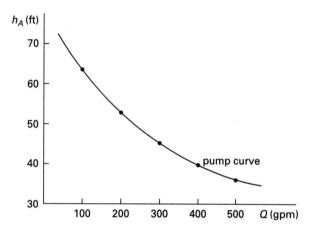

Now construct the system curve. Start by assuming a flow rate of $Q = 300$ gal/min, since this value is in the midrange of the pump characteristic curve. (A flow velocity also could be assumed.)

The flow velocity through the pipe is

$$v = \frac{Q}{A} = \frac{\left(300 \dfrac{\text{gal}}{\text{min}}\right)\left(0.002228 \dfrac{\text{ft}^3\text{-min}}{\text{gal-sec}}\right)}{0.2006 \text{ ft}^2}$$
$$= 3.33 \text{ ft/sec}$$

At 60°F, the kinematic viscosity of the water is

$$\nu = 1.217 \times 10^{-5} \text{ ft}^2/\text{sec}$$

The Reynolds number is

$$\text{Re} = \frac{vD}{\nu} = \frac{\left(3.33 \dfrac{\text{ft}}{\text{sec}}\right)(0.5054 \text{ ft})}{1.217 \times 10^{-5} \dfrac{\text{ft}^2}{\text{sec}}}$$
$$= 1.38 \times 10^5$$

The relative roughness was found previously to be 0.0004. From a Moody friction factor chart for a relative roughness of $\epsilon/D = 0.0004$, and for a Reynolds number of $\text{Re} = 1.38 \times 10^5$, $f = 0.019$.

The equivalent length of pipe is 3000 ft.

The friction head loss for a flow rate of 300 gal/min is

$$h_f = \frac{fL_e v^2}{2Dg} = \frac{(0.019)(3000 \text{ ft})\left(3.33 \dfrac{\text{ft}}{\text{sec}}\right)^2}{(2)(0.5054 \text{ ft})\left(32.2 \dfrac{\text{ft}}{\text{sec}^2}\right)}$$
$$= 19.42 \text{ ft}$$

The velocity head added by the pump is

$$h_v = \frac{v^2}{2g} = \frac{\left(3.33 \dfrac{\text{ft}}{\text{sec}}\right)^2}{(2)\left(32.2 \dfrac{\text{ft}}{\text{sec}^2}\right)}$$
$$= 0.17 \text{ ft}$$

The velocity head is small and could be omitted.

The total head added by the pump at this flow rate is

$$h_{A,300 \text{ gal/min}} = \Delta z + h_f + h_v$$
$$= (20 \text{ ft} + 20 \text{ ft}) + 19.42 \text{ ft} + 0.17 \text{ ft}$$
$$= 59.59 \text{ ft}$$

This provides only a single point on the system curve. The same procedure could be used to obtain a set of system points. However, it is easier to use this point to generate the others.

Make the following assumptions, which are valid for points not too much above and below the original operating point of 300 gal/min: (1) The surface level in the supply reservoir does not change. (2) The friction factor is not influenced greatly by the flow rate. (3) The minor loss, friction loss, and velocity head are proportional to v^2.

Since v is proportional to Q, v^2 will be proportional to Q^2. Then, the general equation for head added by the pump is

$$h_{A,Q} = \Delta z + h_f + h_v$$

$$= (20 \text{ ft} + 20 \text{ ft}) + (19.42 \text{ ft} + 0.17 \text{ ft})\left(\frac{Q}{300 \dfrac{\text{gal}}{\text{min}}}\right)^2$$

$$= 40 \text{ ft} + (19.59 \text{ ft})\left(\frac{Q}{300 \dfrac{\text{gal}}{\text{min}}}\right)^2$$

The system curve is generated by substituting different values of Q.

Q (gpm)	h_A (ft)
100	42.2 ft
200	48.7 ft
300	59.6 ft
400	74.8 ft
500	94.4 ft

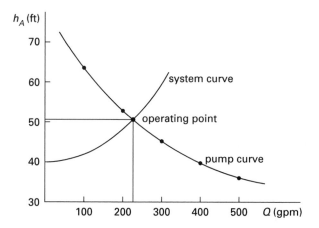

The intersection of the pump characteristic and system curves defines the operating point.

$$Q = 225 \text{ gal/min}$$
$$h_A = 51 \text{ ft}$$

At this point, a better estimate of the flow rate and friction factor used in the system curve generating equation can be determined.

$$Q = 225 \text{ gal/min}$$
$$v = 2.5 \text{ ft/sec}$$
$$\nu = 1.217 \times 10^{-5} \text{ ft}^2/\text{sec}$$
$$\text{Re} = 1.04 \times 10^5$$
$$\epsilon/D = 0.0004$$
$$f = 0.02$$
$$h_f = 11.5 \text{ ft}$$
$$h_v = 0.10 \text{ ft}$$

$$h_{A,225 \text{ gal/min}} = (20 \text{ ft} + 20 \text{ ft}) + 11.5 \text{ ft} + 0.10 \text{ ft}$$
$$= 51.6 \text{ ft}$$

$$h_{A,Q} = 40 \text{ ft} + (11.5 \text{ ft} + 0.10 \text{ ft})\left(\frac{Q}{225 \frac{\text{gal}}{\text{min}}}\right)^2$$
$$= 40 \text{ ft} + (11.6 \text{ ft})\left(\frac{Q}{225 \frac{\text{gal}}{\text{min}}}\right)^2$$

Q (gpm)	h_A (ft)
100	42.3 ft
200	49.2 ft
300	60.6 ft
400	76.7 ft
500	97.3 ft

The curve is not significantly different around the operating point. The revised operating point values are

$$Q = 225 \text{ gal/min}$$
$$h_A = 51.5 \text{ ft}$$

The answer is B.

53. This was determined in the previous question as 51.5 ft.

The answer is B.

54. Schedule-80 pipe has the same outside diameter as schedule-40 pipe, but the inside diameter is smaller. Therefore, the friction loss will be greater in the schedule-80 pipe. Although the same pump curve applies, the system curve is different. The zero flow-head (40 ft) is the same, but the curve is shifted upward. This will shift the operating point upward and to the left. The head added by the pump will increase; the flow rate will decrease.

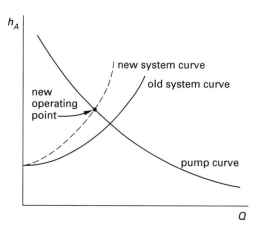

The answer is B.

55. The hydraulic (water) horsepower is approximately

$$\text{WHP} = \frac{h_A Q (\text{SG})}{3956 \frac{\text{gal-ft}}{\text{hp-min}}}$$
$$= \frac{(51.5 \text{ ft})\left(225 \frac{\text{gal}}{\text{min}}\right)(1.0)}{3956 \frac{\text{gal-ft}}{\text{hp-min}}}$$
$$= 2.93 \text{ hp}$$

The answer is C.

56. Since the volumetric flow rate is low, the pump efficiency is probably not much greater than 70%.

$$\text{EHP} = \frac{\text{WHP}}{\eta_{\text{pump}}} = \frac{2.93 \text{ hp}}{0.7}$$
$$= 4.2 \text{ hp}$$

Select a 5 hp motor.

The answer is C.

57. Disregarding slip, the synchronous speed is approximately

$$n_{\text{synchronous}} = \frac{120f}{p} = \frac{(120)(60 \text{ Hz})}{4}$$
$$= 1800 \text{ rpm}$$

The specific speed is approximately

$$n_s = \frac{n\sqrt{Q}}{h_A^{0.75}} = \frac{(1800 \text{ rpm})\sqrt{225 \, \frac{\text{gal}}{\text{min}}}}{(51.5 \text{ ft})^{0.75}}$$
$$= 1404 \text{ rpm}$$

The answer is C.

58. The pump efficiency cannot be determined exactly since the efficiency data was not included with the manufacturer's data. However, generic charts of pump efficiency versus flow rates are common in pump and fluid books. From such a curve in the *Civil Engineering Reference Manual* (Lindeburg) for $Q = 225$ gpm and $n_s = 1400$, the efficiency is approximately 75%.

The answer is C.

59. If the pump casing is unchanged and the impeller size is varied, the flow rate is proportional to the impeller diameter. If the impeller diameter is decreased approximately 10%, the flow rate will decrease approximately 10%.

The answer is D.

60. The suction lift for this installation is high. The maximum theoretical suction lift is 33 ft with a perfect vacuum. Depending on the manufacturer's recommendations, the suction lift probably should not exceed 10 to 15 ft. (A) Using two pumps in parallel will halve the flow rate for each pump, reducing the specific speed and thereby reducing the pump efficiency, without reducing the suction lift. (B) The pump is already operating in the specific speed range appropriate to a radial vane impeller. All other impeller types (e.g., mixed- and axial-flow) are designed for much higher specific speeds.

(C) Constricting the suction line will make the pump more likely to cavitate. (D) Locating the pump at a lower elevation will not decrease the total head that is added. However, it will reduce the large suction lift.

The answer is D.

61. The water will begin to be drawn from the reservoir after 10 hours, when the offsite storage is depleted. By that time, all particles larger than the cutoff size must have settled out. The minimum settling velocity is

$$v_{s,\text{min}} = \frac{s}{t} = \frac{15 \text{ ft}}{10 \text{ hr}}$$
$$= 1.5 \text{ ft/hr}$$

The answer is B.

62. The kinematic viscosity of $40°\text{F}$ water is

$$\nu = 1.664 \times 10^{-5} \text{ ft}^2/\text{sec}$$

The particles are approximately spherical. The Reynolds number is

$$\text{Re} = \frac{v_s D}{\nu} = \frac{\left(1.5 \, \frac{\text{ft}}{\text{hr}}\right)(4 \times 10^{-4} \text{ in})}{\left(3600 \, \frac{\text{sec}}{\text{hr}}\right)\left(12 \, \frac{\text{in}}{\text{ft}}\right)\left(1.664 \times 10^{-5} \, \frac{\text{ft}^2}{\text{sec}}\right)}$$
$$= 8.35 \times 10^{-4}$$

The answer is A.

63. Stokes' law relates the drag coefficient of small spheres in slow motion through a fluid to the settling velocity.

The answer is D.

64. Since $\text{Re} < 1$, Stokes' law is used to calculate the settling velocity.

$$v_{s,\text{actual}} = \frac{D^2 g (\rho_{\text{particle}} - \rho_{\text{water}})}{18 \mu g_c}$$
$$= \frac{D^2 g (\text{SG}_{\text{particle}} - 1)}{18\nu}$$
$$= \frac{\begin{pmatrix} (4 \times 10^{-4} \text{ in})^2 \left(32.2 \, \frac{\text{ft}}{\text{sec}^2}\right) \\ \times (2.65 - 1) \left(3600 \, \frac{\text{sec}}{\text{hr}}\right) \end{pmatrix}}{\left(12 \, \frac{\text{in}}{\text{ft}}\right)^2 (18) \left(1.664 \times 10^{-5} \, \frac{\text{ft}^2}{\text{sec}}\right)}$$
$$= 0.71 \text{ ft/hr}$$

The answer is A.

65. If a particle of the maximum diameter starts at the water surface, the time required to settle out will be

$$t = \frac{s}{v_{s,\text{actual}}} = \frac{15 \text{ ft}}{0.71 \frac{\text{ft}}{\text{hr}}}$$

$$= 21.1 \text{ hr}$$

The answer is D.

66. The geometric slope is

$$S_0 = \frac{\Delta z}{L} = \frac{2 \text{ ft}}{(1 \text{ mi}) \left(5280 \frac{\text{ft}}{\text{mi}}\right)}$$

$$= 0.0003788 \text{ ft/ft}$$

The answer is B.

67. Since the maximum depth of flow is 10 ft and the side slope is 1:2, the horizontal run of each inclined side is 20 ft. The wetted length is found from the Pythagorean theorem.

$$s = \sqrt{(10 \text{ ft})^2 + (20 \text{ ft})^2} = 22.36 \text{ ft}$$

The answer is D.

68. The base dimension, b, was not specified. The base can be calculated from the flow rate. Let w represent the unknown channel width at the surface. Then, the base length is

$$b = w - (2)(20 \text{ ft}) = w - 40 \text{ ft}$$

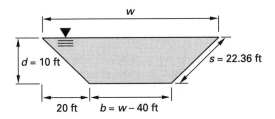

The equivalent diameter of a trapezoidal channel is

$$D_e = \frac{2d(w + b)}{b + 2s}$$

$$= \frac{(2)(10 \text{ ft})(w + w - 40 \text{ ft})}{(w - 40 \text{ ft}) + (2)(22.36 \text{ ft})}$$

$$= \frac{(40 \text{ ft})(w - 20 \text{ ft})}{w + 4.72 \text{ ft}}$$

The hydraulic radius of a trapezoidal channel is

$$R = \frac{D_e}{4} = \frac{\dfrac{(40 \text{ ft})(w - 20 \text{ ft})}{w + 4.72 \text{ ft}}}{4}$$

$$= \frac{(10 \text{ ft})(w - 20 \text{ ft})}{w + 4.72 \text{ ft}}$$

The area in flow is

$$A = (2) \left(\tfrac{1}{2}\right)(20 \text{ ft})(10 \text{ ft}) + (w - 40 \text{ ft})(10 \text{ ft})$$

$$= 200 \text{ ft}^2 + (10 \text{ ft})w - 400 \text{ ft}^2$$

$$= (10 \text{ ft})w - 200 \text{ ft}^2$$

Use the Chezy-Manning equation. Since the flow is uniform, the hydraulic gradient, S, is the geometric slope, S_0.

$$Q = \left(\frac{1.49}{n}\right) A R^{\frac{2}{3}} \sqrt{S}$$

$$2200 \frac{\text{ft}^3}{\text{sec}} = \left(\frac{1.49}{0.013}\right)\left((10 \text{ ft})w - 200 \text{ ft}^2\right)$$

$$\times \left(\frac{(10 \text{ ft})(w - 20 \text{ ft})}{w + 4.72 \text{ ft}}\right)^{\frac{2}{3}} \sqrt{0.0003788}$$

Simplifying,

$$986.2 = \left(\frac{(10w - 200)^5}{(w + 4.72)^2}\right)^{\frac{1}{3}}$$

Using an equation solver or by trial and error, $w = 51.3$ ft.

The answer is C.

69. Substitute the top width into the equation for equivalent diameter.

$$D_e = \frac{(40 \text{ ft})(w - 20 \text{ ft})}{w + 4.72 \text{ ft}}$$

$$= \frac{(40 \text{ ft})(51.3 \text{ ft} - 20 \text{ ft})}{51.3 \text{ ft} + 4.72 \text{ ft}}$$

$$= 22.35 \text{ ft}$$

The answer is A.

70. The most efficient trapezoidal channel has sides inclined 60° from the horizontal. This is a ratio of 1.732:1.0, the same as 1.0:0.577.

The answer is B.

71. The Manning roughness coefficient of typically formed concrete can vary from approximately 0.009 for very smooth concrete to 0.017 for very rough concrete. The stilling basin is probably neither excessively smooth nor rough, so a midrange value of 0.014 is appropriate.

The answer is B.

72. Contraction of water jets is primarily caused by momentum effects. Contraction can be reduced or eliminated by careful attention to the design of the orifice geometry.

The answer is D.

73. Flow changes from being supercritical (fast and shallow) to being subcritical (slow and deep) in a hydraulic jump. Since it is known that the depth increases in a hydraulic jump somewhere downstream, the original flow must be supercritical (not critical or subcritical). Since the depth gradually increases to the point where the hydraulic jump occurs, the velocity will also be gradually decreasing. Thus, the flow is retarded.

The answer is D.

74. The hydraulic radius is the area in flow divided by the wetted perimeter.

$$R = \frac{A}{P}$$

The area in flow is

$$A = dw = (4.08 \text{ ft})w$$

The wetted perimeter for a rectangular open channel would normally be calculated as

$$P = w + 2d$$

Although the width is unknown, the basin is said to be wide. For wide channels, $2d \ll w$, so that

$$P \approx w$$

$$R = \frac{A}{P} \approx \frac{(4.08 \text{ ft})w}{w} = 4.08 \text{ ft}$$

The answer is D.

75. Since depth and velocity are inversely proportional, the velocity can be scaled down from the point of minimum depth.

$$v_0 d_0 = v_1 d_1$$

$$\left(62 \ \frac{\text{ft}}{\text{sec}}\right)(4.08 \text{ ft}) = v_1(4.57 \text{ ft})$$

$$v_1 = 55.35 \text{ ft/sec}$$

The answer is B.

76. The depths immediately before and after a hydraulic jump are the conjugate depths, d_1 and d_2. One can be calculated from the other.

$$d_2 = -\left(\frac{d_1}{2}\right) + \sqrt{\frac{2v_1^2 d_1}{g} + \frac{d_1^2}{4}}$$

$$= -\left(\frac{4.57 \text{ ft}}{2}\right)$$

$$+ \sqrt{\frac{(2)\left(55.35 \ \dfrac{\text{ft}}{\text{sec}}\right)^2 (4.57 \text{ ft})}{32.2 \ \dfrac{\text{ft}}{\text{sec}^2}} + \frac{(4.57 \text{ ft})^2}{4}}$$

$$= 27.29 \text{ ft}$$

The answer is D.

77. The velocity after the jump is

$$v_2 = \left(\frac{d_0}{d_2}\right)v_0 = \left(\frac{4.08 \text{ ft}}{25 \text{ ft}}\right)\left(62 \ \frac{\text{ft}}{\text{sec}}\right)$$

$$= 10.1 \text{ ft/sec}$$

The answer is C.

78. The average velocity between the locations of the minimum depth and the hydraulic jump is

$$v_{\text{ave}} = \frac{v_0 + v_1}{2} = \frac{62 \ \dfrac{\text{ft}}{\text{sec}} + 55.35 \ \dfrac{\text{ft}}{\text{sec}}}{2}$$

$$= 58.68 \text{ ft/sec}$$

The average hydraulic radius between these two points is

$$R_{\text{ave}} = \frac{A}{P} \approx \frac{d_{\text{ave}}w}{w} = d_{\text{ave}} = \frac{d_0 + d_1}{2}$$

$$= \frac{4.08 \text{ ft} + 4.57 \text{ ft}}{2}$$

$$= 4.325 \text{ ft}$$

The stilling basin is horizontal, so the geometric gradient, S_0, is zero. The average energy gradient is the slope of the energy line.

$$S_{\text{ave}} = \left(\frac{n v_{\text{ave}}}{(1.49)(R_{\text{ave}})^{\frac{2}{3}}}\right)^2$$

$$= \left(\frac{(0.013)\left(58.68 \ \dfrac{\text{ft}}{\text{sec}}\right)}{(1.49)(4.325 \text{ ft})^{\frac{2}{3}}}\right)^2$$

$$= 0.0372$$

The answer is B.

79. The distance between the two points is

$$L = \frac{d_0 + \frac{v_0^2}{2g} - \left(d_1 + \frac{v_1^2}{2g}\right)}{S - S_0}$$

$$= \frac{\left(4.08 \text{ ft} + \frac{\left(62 \frac{\text{ft}}{\text{sec}}\right)^2}{(2)\left(32.2 \frac{\text{ft}}{\text{sec}^2}\right)}\right) - \left(4.57 \text{ ft} + \frac{\left(55.35 \frac{\text{ft}}{\text{sec}}\right)^2}{(2)\left(32.2 \frac{\text{ft}}{\text{sec}^2}\right)}\right)}{0.0372 - 0}$$

$$= 312.6 \text{ ft}$$

The answer is C.

80. The specific energy loss is the difference of the specific energies before and after the jump.

$$\Delta E = E_1 - E_2 = d_1 + \frac{v_1^2}{2g} - d_2 - \frac{v_2^2}{2g}$$

$$= \left(4.57 \text{ ft} + \frac{\left(55.35 \frac{\text{ft}}{\text{sec}}\right)^2}{(2)\left(32.2 \frac{\text{ft}}{\text{sec}^2}\right)}\right) - \left(27.29 \text{ ft} + \frac{\left(10.1 \frac{\text{ft}}{\text{sec}}\right)^2}{(2)\left(32.2 \frac{\text{ft}}{\text{sec}}^2\right)}\right)$$

$$= 23.3 \text{ ft}$$

The answer is C.

Solutions
Structural

81. Assume a beam depth of 24 in. Reinforced concrete has a unit weight of approximately 150 lbf/ft^3. The self-weight will be

$$w_{\text{beam}} = \gamma A = \gamma bh$$

$$= \frac{\left(150\ \dfrac{\text{lbf}}{\text{ft}^3}\right)(18\ \text{in})(24\ \text{in})}{\left(12\ \dfrac{\text{in}}{\text{ft}}\right)^2}$$

$$= 450\ \text{lbf/ft}$$

The total dead load per foot is

$$w_d = 1500\ \frac{\text{lbf}}{\text{ft}} + 450\ \frac{\text{lbf}}{\text{ft}}$$

$$= 1950\ \text{lbf/ft}$$

The bending moment at the built-in end due to the dead load is

$$M_d = \frac{w_d L^2}{2} = \frac{\left(1950\ \dfrac{\text{lbf}}{\text{ft}}\right)(18\ \text{ft})^2}{2}$$

$$= 315{,}900\ \text{ft-lbf}$$

The moment at the built-in end due to the live load at the tip is

$$M_l = PL = (10{,}000\ \text{lbf})(18\ \text{ft})$$

$$= 180{,}000\ \text{ft-lbf}$$

The ultimate moment is

$$M_u = 1.4M_d + 1.7M_l$$

$$= (1.4)(315{,}900\ \text{ft-lbf}) + (1.7)(180{,}000\ \text{ft-lbf})$$

$$= 748{,}260\ \text{ft-lbf}$$

The answer is C.

82. The minimum reinforcement ratio is

$$\rho_{\min} = \frac{200}{f_y} = \frac{200}{60{,}000\ \dfrac{\text{lbf}}{\text{in}^2}}$$

$$= 0.00333$$

The answer is B.

83. Since $f_c' = 3000$ psi, $\beta_1 = 0.85$. The balanced reinforcement ratio is

$$\rho_b = \left((0.85)\left(\frac{\beta_1 f_c'}{f_y}\right)\right)\left(\frac{87{,}000}{87{,}000 + f_y}\right)$$

$$= \left(\frac{(0.85)(0.85)\left(3000\ \dfrac{\text{lbf}}{\text{in}^2}\right)}{60{,}000\ \dfrac{\text{lbf}}{\text{in}^2}}\right)\left(\frac{87{,}000}{87{,}000 + 60{,}000}\right)$$

$$= 0.0214$$

The maximum allowable reinforcement ratio is 75% of the balanced value. Half of the allowable value is

$$\rho = (0.5)(0.75)(0.0214)$$

$$= 0.00803$$

The answer is A.

84. The ultimate beam strength required is

$$M_n = \frac{M_u}{\phi} = \rho bd^2 f_y\left(1 - \frac{\rho f_y}{1.7 f_c'}\right)$$

Solving for the beam depth,

$$d = \sqrt{\frac{M_u}{\phi \rho b f_y\left(1 - \dfrac{\rho f_y}{1.7 f_c'}\right)}}$$

$$= \sqrt{\frac{(748{,}260\ \text{ft-lbf})\left(12\ \dfrac{\text{in}}{\text{ft}}\right)}{(0.90)(0.00803)(18\ \text{in})\left(60{,}000\ \dfrac{\text{lbf}}{\text{in}^2}\right) \times \left(1 - \dfrac{(0.00803)\left(60{,}000\ \dfrac{\text{lbf}}{\text{in}^2}\right)}{(1.7)\left(3000\ \dfrac{\text{lbf}}{\text{in}^2}\right)}\right)}}$$

$$= 35.64\ \text{in}$$

This is considerably larger than the beam size of 24 in initially assumed. Use a second iteration. Allowing approximately 4 in for cover and stirrups, try $h = 40$ in.

$$w_{\text{beam}} = \gamma A = \gamma bh$$

$$= \frac{\left(150 \ \frac{\text{lbf}}{\text{ft}^3}\right)(18 \text{ in})(40 \text{ in})}{\left(12 \ \frac{\text{in}}{\text{ft}}\right)^2}$$

$$= 750 \text{ lbf/ft}$$

The total dead load per foot is

$$w_d = 1500 \ \frac{\text{lbf}}{\text{ft}} + 750 \ \frac{\text{lbf}}{\text{ft}}$$

$$= 2250 \text{ lbf/ft}$$

The bending moment at the built-in end due to the dead load is

$$M_d = \frac{w_d L^2}{2} = \frac{\left(2250 \ \frac{\text{lbf}}{\text{ft}}\right)(18 \text{ ft})^2}{2}$$

$$= 364{,}500 \text{ ft-lbf}$$

$$M_u = 1.4 M_d + 1.7 M_l$$

$$= (1.4)(364{,}500 \text{ ft-lbf}) + (1.7)(180{,}000 \text{ ft-lbf})$$

$$= 816{,}300 \text{ ft-lbf}$$

Solving for the beam depth,

$$d = \sqrt{\frac{M_u}{\phi \rho b f_y \left(1 - \frac{\rho f_y}{1.7 f'_c}\right)}}$$

$$= \sqrt{\frac{(816{,}300 \text{ ft-lbf})\left(12 \ \frac{\text{in}}{\text{ft}}\right)}{(0.90)(0.00803)(18 \text{ in})\left(60{,}000 \ \frac{\text{lbf}}{\text{in}^2}\right)} \times \left(1 - \frac{(0.00803)\left(60{,}000 \ \frac{\text{lbf}}{\text{in}^2}\right)}{(1.7)\left(3000 \ \frac{\text{lbf}}{\text{in}^2}\right)}\right)}$$

$$= 37.23 \text{ in} \quad [\text{use } 37 \text{ in}]$$

The answer is C.

85. The required steel area is

$$A_s = \rho bd = (0.00803)(18 \text{ in})(37 \text{ in})$$

$$= 5.35 \text{ in}^2$$

The cross-sectional area of a no. 9 bar is 1.00 in^2. Six bars will be needed to provide 5.35 in^2.

The answer is A.

86. All of the steel is placed above the neutral axis because the beam is bent with tension in the top fibers.

The answer is D.

87. Since the beam is not exposed to earth or weather, only $1\frac{1}{2}$ in of concrete cover is required. A no. 9 bar has a diameter of 1.128 in, and a no. 3 stirrup has a diameter of 0.375 in. The minimum beam height is

$$h = d + \frac{d_b}{2} + d_s + \text{cover}$$

$$= 37 \text{ in} + \frac{1.128 \text{ in}}{2} + 0.375 \text{ in} + 1.5 \text{ in}$$

$$= 39.4 \text{ in} \quad [\text{use } 39.5 \text{ in}]$$

The answer is C.

88. The cracked moment of inertia assumes that the beam is cracked above its neutral axis. The cracked section does not contribute to the deflection resistance. The location of the neutral axis depends on the modular ratio.

The concrete's modulus of elasticity is

$$E_c = 57{,}000\sqrt{f'_c}$$

$$= 57{,}000\sqrt{3000 \ \frac{\text{lbf}}{\text{in}^2}}$$

$$= 3.12 \times 10^6 \ \frac{\text{lbf}}{\text{in}^2}$$

The modular ratio is

$$n = \frac{E_s}{E_c} = \frac{29 \times 10^6 \ \frac{\text{lbf}}{\text{in}^2}}{3.12 \times 10^6 \ \frac{\text{lbf}}{\text{in}^2}}$$

$$= 9.29$$

Refer to the following diagram. The actual steel area of 6 in^2 is increased by the modular ratio. The transformed area of steel (the area of equivalent concrete) is

$$A'_s = nA_s = (9.29)(6 \text{ in}^2) = 55.74 \text{ in}^2$$

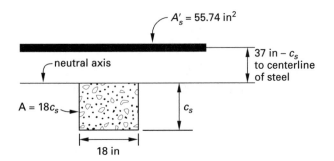

The neutral axis can be located by a number of methods. A direct calculation is the most expedient.

$$
\begin{aligned}
c_s &= \left(\frac{nA_s}{b}\right)\left(\sqrt{1 + \frac{2bd}{nA_s}} - 1\right) \\
&= \left(\frac{55.74 \text{ in}^2}{18 \text{ in}}\right)\left(\sqrt{1 + \frac{(2)(18 \text{ in})(37 \text{ in})}{55.74 \text{ in}^2}} - 1\right) \\
&= 12.35 \text{ in}
\end{aligned}
$$

The cracked moment of inertia is

$$
\begin{aligned}
I_{\text{cr}} &= \frac{bc_s^3}{3} + nA_s(d - c_s)^2 \\
&= \frac{(18 \text{ in})(12.35 \text{ in})^3}{3} \\
&\quad + (55.74 \text{ in}^2)(37 \text{ in} - 12.35 \text{ in})^2 \\
&= 45{,}171 \text{ in}^4
\end{aligned}
$$

The answer is D.

89. The gross moment of inertia disregards any cracking in the beam.

$$
\begin{aligned}
I_g &= \frac{bh^3}{12} = \frac{(18 \text{ in})(39.5 \text{ in})^3}{12} \\
&= 92{,}445 \text{ in}^4
\end{aligned}
$$

The modulus of rupture is

$$
\begin{aligned}
f_r &= 7.5\sqrt{f_c'} = 7.5\sqrt{3000 \dfrac{\text{lbf}}{\text{in}^2}} \\
&= 410.8 \text{ lbf/in}^2
\end{aligned}
$$

The distance, y_t, between the centroid of the gross section (neglecting reinforcement) and the extreme tension fiber is half of the beam height.

$$
\begin{aligned}
y_t &= \frac{h}{2} = \frac{39.5 \text{ in}}{2} \\
&= 19.75 \text{ in}
\end{aligned}
$$

The moment that would first crack the concrete is

$$
\begin{aligned}
M_{\text{cr}} &= \frac{f_r I_g}{y_t} \\
&= \frac{\left(410.8 \dfrac{\text{lbf}}{\text{in}^2}\right)(92{,}445 \text{ in}^4)}{19.75 \text{ in}} \\
&= 1.923 \times 10^6 \text{ in-lbf}
\end{aligned}
$$

$$
w_{\text{beam}} = \gamma A = \gamma bh
$$

$$
\begin{aligned}
&= \frac{\left(150 \dfrac{\text{lbf}}{\text{ft}^3}\right)(18 \text{ in})(39.5 \text{ in})}{\left(12 \dfrac{\text{in}}{\text{ft}}\right)^2} \\
&= 741 \text{ lbf/ft}
\end{aligned}
$$

The total dead load per foot is

$$
\begin{aligned}
w_d &= 1500 \frac{\text{lbf}}{\text{ft}} + 741 \frac{\text{lbf}}{\text{ft}} \\
&= 2241 \text{ lbf/ft}
\end{aligned}
$$

The bending moment at the built-in end due to the dead load is

$$
\begin{aligned}
M_d &= \frac{w_d L^2}{2} = \frac{\left(2241 \dfrac{\text{lbf}}{\text{ft}}\right)(18 \text{ ft})^2}{2} \\
&= 363{,}042 \text{ ft-lbf}
\end{aligned}
$$

The actual (unfactored) moment acting on the beam at the support is

$$
\begin{aligned}
M_a &= M_d + M_l \\
&= (363{,}042 \text{ ft-lbf} + 180{,}000 \text{ ft-lbf})\left(12 \frac{\text{in}}{\text{ft}}\right) \\
&= 6.517 \times 10^6 \text{ in-lbf}
\end{aligned}
$$

The ratio of the cracking to actual moments is

$$
\begin{aligned}
\frac{M_{\text{cr}}}{M_a} &= \frac{1.923 \times 10^6 \text{ in-lbf}}{6.517 \times 10^6 \text{ in-lbf}} \\
&= 0.295
\end{aligned}
$$

The effective moment of inertia is

$$
\begin{aligned}
I_e &= \left(\frac{M_{\text{cr}}}{M_a}\right)^3 I_g + \left(1 - \left(\frac{M_{\text{cr}}}{M_a}\right)^3\right) I_{\text{cr}} \\
&= (0.295)^3(92{,}445 \text{ in}^4) + \left(1 - (0.295)^3\right)(45{,}171 \text{ in}^4) \\
&= 46{,}385 \text{ in}^4
\end{aligned}
$$

The answer is C.

90. The effective moment of inertia is used with standard beam deflection equations to calculate the instantaneous deflection. Since there are two loadings (uniform and concentrated), the total deflection is the superposition of two separate terms.

$$
w_d = \frac{2241 \dfrac{\text{lbf}}{\text{ft}}}{12 \dfrac{\text{in}}{\text{ft}}} = 186.8 \text{ lbf/in}
$$

$$P_{tip} = 10,000 \text{ lbf}$$

$$L = (18 \text{ ft})\left(12 \, \frac{\text{in}}{\text{ft}}\right) = 216 \text{ in}$$

$$\Delta_i = \Delta_w + \Delta_P$$

$$= \frac{w_d L^4}{8EI_e} + \frac{PL^3}{3EI_e}$$

$$= \frac{\left(186.8 \, \frac{\text{lbf}}{\text{in}}\right)(216 \text{ in})^4}{(8)\left(3.12 \times 10^6 \, \frac{\text{lbf}}{\text{in}^2}\right)(46,385 \text{ in}^4)}$$

$$+ \frac{(10,000 \text{ lbf})(216 \text{ in})^3}{(3)\left(3.12 \times 10^6 \, \frac{\text{lbf}}{\text{in}^2}\right)(46,385 \text{ in}^4)}$$

$$= 0.583 \text{ in}$$

The answer is B.

91. The total load on the roof is the sum of the dead and live loads.

$$w = w_d + w_l = 20 \, \frac{\text{lbf}}{\text{ft}^2} + 25 \, \frac{\text{lbf}}{\text{ft}^2}$$

$$= 45 \text{ lbf/ft}^2$$

Except for the end beams, each interior beam supports a tributary area of

$$A = (10 \text{ ft})(30 \text{ ft}) = 300 \text{ ft}^2$$

The roof load supported by each interior glulam beam is

$$W_{\text{roof}} = wA = \left(45 \, \frac{\text{lbf}}{\text{ft}^2}\right)(300 \text{ ft}^2)$$

$$= 13,500 \text{ lbf}$$

In addition, some allowance should be made for beam weight. The beam size is not known at this point. Calculating a self-weight from the beam volume and unit weight is as much of an estimate as merely allowing 25 lbf/ft^2.

$$W_{\text{beam}} = (30 \text{ ft})\left(25 \, \frac{\text{lbf}}{\text{ft}}\right)$$

$$= 750 \text{ lbf}$$

The total load supported by the beam is

$$W_{\text{total}} = W_{\text{roof}} + W_{\text{beam}} = 13,500 \text{ lbf} + 750 \text{ lbf}$$

$$= 14,250 \text{ lbf}$$

The answer is D.

92. "24F" means that the maximum allowable tensile bending stress in the tensile region of the beam is 2400 psi, exclusive of size factors.

The answer is A.

93. "V4" means that the beam has been constructed from visually graded lumber, not pieces that have been individually tested for strength.

The answer is B.

94. These glulam beams are constructed from nominal 2×6 members. The actual width of a 6 in member is approximately $5^1/_2$ in. After finishing, the glulam beam will be approximately $5^1/_8$ in wide. Other widths are always available on special order, but $5^1/_8$ in widths are standard.

The answer is A.

95. Glulams and their installation methods are not inherently more resistant to buckling or more complex than beams of any other material. Nor is the complexity of the buckling checking so excessive as to make omitting it an unreasonable request. In this design, the tendency to buckle must be resisted from other roof framing.

The answer is B.

96. The uniform load supported per foot is

$$w = \frac{W}{L} = \frac{14,250 \text{ lbf}}{30 \text{ ft}}$$

$$= 475 \text{ lbf/ft}$$

The maximum moment on the beam is

$$M_{\max} = \frac{wL^2}{8} = \frac{\left(475 \, \frac{\text{lbf}}{\text{ft}}\right)(30 \text{ ft})^2}{8}$$

$$= 53,438 \text{ ft-lbf}$$

The allowable bending stress is 2400 psi. (No factor of safety is required.) The bending stress is

$$f = \frac{Mc}{I} = \frac{M\left(\dfrac{d}{2}\right)}{\dfrac{bd^3}{12}} = \frac{6M}{bd^2}$$

The required beam depth is

$$d = \sqrt{\frac{6M}{bf}}$$

$$= \sqrt{\frac{(6)(53,438 \text{ ft-lbf})\left(12 \, \dfrac{\text{in}}{\text{ft}}\right)}{(5.125 \text{ in})\left(2400 \, \dfrac{\text{lbf}}{\text{in}^2}\right)}}$$

$$= 17.7 \text{ in} \quad [\text{use 18 in}]$$

At this point, the assumption of beam weight can be checked. The unit weight of Douglas fir is approximately 36 lbf/ft^3. The weight per foot of beam is

$$w_{\text{beam}} = \gamma A$$

$$= \frac{\left(36 \, \frac{\text{lbf}}{\text{ft}^3}\right)(5.125 \text{ in})(18 \text{ in})}{\left(12 \, \frac{\text{in}}{\text{ft}}\right)^2}$$

$$= 23.1 \text{ lbf/ft}$$

This is close enough to the initial assumption of 25 lbf/ft. No adjustments are required.

Since the depth is more than 12 in, a volume factor, C_v, must be included. Refer to the 1997 NDS ASD Supplement "Structural Glued Laminated Timber," Sec. 4.9.

$$K_L = 1.00 \quad [\text{Table 4.4, uniform load}]$$

$$C_{gd} = \left(\frac{12 \text{ in}}{18 \text{ in}}\right)^{0.10} = 0.960 \quad [\text{Table 4.5, Western Species}]$$

$$C_{gb} = 1.00 \quad [5^1\!/\!8 \text{ in width}]$$

$$C_{gl} = \left(\frac{21 \text{ ft}}{30 \text{ ft}}\right)^{0.10} = 0.965 \quad [\text{Table 4.5, Western Species}]$$

$$C_v = K_L C_{gd} C_{gb} C_{gl} = (1.00)(0.960)(1.00)(0.965)$$

$$= 0.926$$

Recalculate the depth.

$$d = \sqrt{\frac{6M}{bC_F f}}$$

$$= \sqrt{\frac{(6)(53{,}438 \text{ ft-lbf})\left(12 \, \frac{\text{in}}{\text{ft}}\right)}{(5.125 \text{ in})(0.926)\left(2400 \, \frac{\text{lbf}}{\text{in}^2}\right)}}$$

$$= 18.38 \text{ in}$$

This is essentially the same as originally calculated. An argument for accepting the slight overstress and avoiding the cost of an additional layer might be valid. However, strictly, a depth increase is called for.

Each layer contributes $1^1\!/\!2$ in to the glulam beam depth. In order to obtain the required depth, 13 layers are required. This requires 12 glue laminations.

The answer is C.

97. From the 1997 NDS ASD Supplement, "Structural Glued Laminated Timber," Table 3.1, for 24F-V4 (DF/DF), the maximum compression stress perpendicular to the grain is 650 psi.

The answer is B.

98. This glulam has a camber and is meant to be installed camber-up. When the intended compression zone is stressed in tension, the allowable extreme bending stress is reduced. From the 1997 NDS ASD Supplement, "Structural Glued Laminated Timber," Table 3.1, for 24F-V4 (DF/DF), the allowable stress is 1850 psi.

The answer is C.

99. From the 1997 NDS ASD Supplement, "Structural Glued Laminated Timber," Table 3.1, for 24F-V4 (DF/DF), the maximum allowable shear stress parallel to the grain is 240 psi. Without the special tensile laminations (note 7), this value must be reduced 25%.

$$F_{vx} = (1 - 0.25)(240 \text{ psi}) = 180 \text{ psi}$$

The answer is A.

100. From the 1997 NDS ASD Supplement, "Structural Glued Laminated Timber," Table 3.1, for 24F-V4 (DF/DF), the modulus of elasticity is

$$E = 1.8 \times 10^6 \text{ lbf/in}^2$$

Using the actual (not nominal) dimensions, the moment of inertia is

$$I = \frac{bd^3}{12} = \frac{(5.125 \text{ in})(18 \text{ in})^3}{12}$$

$$= 2491 \text{ in}^4$$

For a uniformly loaded, simply supported beam, the midpoint deflection is

$$y = \frac{5wL^4}{384EI}$$

$$= \frac{(5)\left(475 \, \frac{\text{lbf}}{\text{ft}}\right)\left((30 \text{ ft})\left(12 \, \frac{\text{in}}{\text{ft}}\right)\right)^4}{(384)\left(1.8 \times 10^6 \, \frac{\text{lbf}}{\text{in}^2}\right)(2491 \text{ in}^4)\left(12 \, \frac{\text{in}}{\text{ft}}\right)}$$

$$= 1.93 \text{ in}$$

The answer is D.

101. Since the bridge is 20 ft wide and 90 ft long, the total vertical load carried by each parallel side of the bridge is

$$F_{\text{side}} = \frac{\left(20 \, \frac{\text{lbf}}{\text{ft}^2}\right)(20 \text{ ft})(90 \text{ ft})}{2 \text{ sides}}$$

$$= 18{,}000 \text{ lbf}$$

Since the two hinge-connected trusses are symmetrical, each truss carries half of this vertical load.

$$R_{A,y} = \frac{F_{side}}{2} = \frac{18,000 \text{ lbf}}{2 \text{ trusses}}$$
$$= 9000 \text{ lbf}$$

The answer is B.

102. The distance between points A and C is 15 ft. The vertical load between points A and C is

$$\frac{\left(20 \frac{\text{lbf}}{\text{ft}^2}\right)(20 \text{ ft})(15 \text{ ft})}{2 \text{ sides}} = 3000 \text{ lbf}$$

This vertical load is divided evenly between points A and C. The contribution of section AC to point C is

$$\frac{3000 \text{ lbf}}{2} = 1500 \text{ lbf}$$

Similarly, the load between points C and E is also 3000 lbf. This divided evenly between points C and E. The contribution of section CE to point C is 1500 lbf.

The total load transmitted to one side at point C is

$$C_y = 1500 \text{ lbf} + 1500 \text{ lbf} = 3000 \text{ lbf}$$

The answer is B.

103. The distance between points E and G is 30 ft. The vertical load between points E and G is

$$\frac{\left(20 \frac{\text{lbf}}{\text{ft}^2}\right)(20 \text{ ft})(30 \text{ ft})}{2 \text{ sides}} = 6000 \text{ lbf}$$

This vertical load is divided evenly between points E and G. The contribution of section EG to point E is

$$\frac{6000 \text{ lbf}}{2} = 3000 \text{ lbf}$$

The contribution of section CE to point E is 1500 lbf.

The total load transmitted to one side at point E is

$$E_y = 3000 \text{ lbf} + 1500 \text{ lbf} = 4500 \text{ lbf}$$

The answer is C.

104. Assume clockwise moments to be positive. Sum moments about point F.

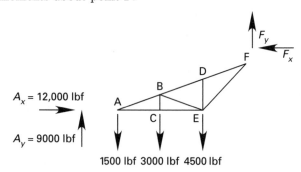

$$\sum M_F = -(4500 \text{ lbf})(15 \text{ ft}) - (3000 \text{ lbf})(30 \text{ ft})$$
$$- (1500 \text{ lbf})(45 \text{ ft}) + (9000 \text{ lbf})(45 \text{ ft})$$
$$- A_x(15 \text{ ft})$$
$$= 0$$
$$A_x = 12,000 \text{ lbf}$$

The answer is D.

105. Since the truss height increases to 15 ft in three equal steps, the length of member BC is 5 ft. The length of member AB is calculated from the Pythagorean theorem.

$$L_{AB} = \sqrt{L_{AC}^2 + L_{BC}^2}$$
$$= \sqrt{(15 \text{ ft})^2 + (5 \text{ ft})^2}$$
$$= 15.81 \text{ ft}$$

The net vertical load at point A is

$$A_y = R_{A,y} - F_A = 9000 \text{ lbf} - 1500 \text{ lbf}$$
$$= 7500 \text{ lbf}$$

Since member AC is horizontal, the entire vertical loading at point A is carried by the vertical force component in member AB. For similarity of shape and force components, the force in member AB is

$$AB = \left(\frac{15.81 \text{ ft}}{5 \text{ ft}}\right)(7500 \text{ lbf}) = 23,715 \text{ lbf}$$

The answer is D.

106. Member DE cannot support a load with a horizontal component. Therefore, the horizontal component of the load in member BD must pass through unchanged to member DF. This means that the vertical components of force are the same in members BD and DF, since these members have the same slopes and lengths. Therefore, DE contributes nothing at point D. Member DE is a zero-force member.

The answer is A.

107. The horizontal component of force in member AB is

$$AB_x = \left(\frac{15 \text{ ft}}{15.81 \text{ ft}}\right)(23{,}715 \text{ lbf}) = 22{,}500 \text{ lbf}$$

Since member AC is horizontal, it can only support a horizontal loading. The force in member AC is

$$AC = AB_x - A_x = 22{,}500 \text{ lbf} - 12{,}000 \text{ lbf}$$
$$= 10{,}500 \text{ lbf}$$

The answer is C.

108. Even though the truss members are considered to be pin-connected for the purposes of force analysis, they are actually welded. They are not pin-connected from the standpoint of the AISC *Manual of Steel Construction*. From Sec. D1, the allowable stress on the gross area of the steel tensile members is

$$F_t = 0.6F_y = (0.6)(36 \text{ ksi})$$
$$= 21.6 \text{ ksi}$$

The answer is C.

109. The maximum allowable tensile stress of 21.6 ksi is permitted on the effective area as designed in Sec. B3. For welded connections, the effective, nominal, and gross areas are identical. The required cross-sectional area is

$$A_g = \frac{AC}{F_t} = \frac{10.5 \text{ kips}}{21.6 \dfrac{\text{kips}}{\text{in}^2}}$$
$$= 0.49 \text{ in}^2$$

The lightest W 8 section is W 8 × 10, which has a cross-sectional area of 2.96 in^2. No W 8 section has a smaller area.

Check the slenderness ratio. The maximum slenderness ratio for main members in tension is 300 [Sec. B7]. The W 8 × 10 has a least radius of gyration of $r_y = 0.841$ in. The slenderness ratio in tension is

$$SR = \frac{L}{r_y} = \frac{(15 \text{ ft})\left(12 \dfrac{\text{in}}{\text{ft}}\right)}{0.841 \text{ in}}$$
$$= 214$$

Since 214 < 300, the W 8 × 10 is satisfactory.

The answer is A.

110. Since all loads are axial, the end connections are made in a manner that allows the ends to rotate (i.e., the ends are "pinned"). The end-restraint coefficient is $K = 1$. The effective length of the member is

$$L_{e,\text{AB}} = KL_{\text{AB}} = (1)(15.81 \text{ ft})\left(12 \dfrac{\text{in}}{\text{ft}}\right)$$
$$= 189.7 \text{ in} \quad [\text{use } 190 \text{ in}]$$

Member AB does not support bending moments and can be sized for pure compression. Member AB carries a load of 23.7 kips. From the AISC Column Tables, W 8 × 24 is the smallest W 8 section, with a capacity of about 81 kips. Although satisfactory, this section has too much capacity. A smaller W 8 section (not found in the Column Tables) should be checked.

Try W 8 × 18 with a minimum radius of gyration of 1.23 in and a cross-sectional area of 5.26 in^2. The axial compressive stress in the member is

$$f_a = \frac{F}{A} = \frac{23.7 \text{ kips}}{5.26 \text{ in}^2}$$
$$= 4.5 \text{ ksi}$$

The maximum slenderness ratio is

$$SR = \frac{KL}{r} = \frac{(1)(15.81 \text{ ft})\left(12 \dfrac{\text{in}}{\text{ft}}\right)}{1.23 \text{ in}}$$
$$= 154$$

From the AISC *Specification* Allowable Stress Table [Table C-36 in the Columns section], the maximum compressive stress permitted for this slenderness ratio is 6.3 ksi. Since 4.5 ksi < 6.3 ksi, this member is satisfactory.

The answer is B.

111.
$$V = \frac{m}{\rho} = \frac{m}{(\text{SG})\rho_w}$$
$$= \frac{517 \text{ lbm}}{(3.15)\left(62.4 \dfrac{\text{lbm}}{\text{ft}^3}\right)}$$
$$= 2.63 \text{ ft}^3$$

The answer is B.

112.
$$m = V\rho = V(\text{SG})\rho_w$$
$$= (10.8 \text{ ft}^3)(2.65)\left(62.4 \dfrac{\text{lbm}}{\text{ft}^3}\right)$$
$$= 1786 \text{ lbm}$$

The answer is D.

113.
$$\text{SG} = \frac{m}{V\rho_w} = \frac{2679 \text{ lbm}}{(16.0 \text{ ft}^3)\left(62.4 \dfrac{\text{lbm}}{\text{ft}^3}\right)}$$
$$= 2.68$$

The answer is C.

114. The total volume is

$$V_t = V_{\text{cement}} + V_{\text{fine}} + V_{\text{coarse}} + V_{\text{water}}$$
$$= 2.63 \text{ ft}^3 + 10.8 \text{ ft}^3 + 16.0 \text{ ft}^3 + 3.72 \text{ ft}^3$$
$$= 33.15 \text{ ft}^3$$

The answer is C.

115. Scaling can be the result of several factors. Usually, poor finishing or inadequate air entrainment are the cause.

The answer is D.

116. The 6 in by 12 in cylinder is typically cast in the field as a means of determining whether the concrete "meets strength" (i.e., has a compressive strength equal to or greater than specified).

$$f_c' = \frac{F}{A} = \frac{100{,}000 \text{ lbf}}{\pi \left(\dfrac{6 \text{ in}}{2} \right)^2}$$
$$= 3537 \text{ psi}$$

The answer is D.

117. The maximum amount of the accelerator is

$$m = (0.02)(8 \text{ yd}^3) \left(6 \, \frac{\text{bags}}{\text{yd}^3} \right) \left(94 \, \frac{\text{lbm}}{\text{bag}} \right)$$
$$= 90.24 \text{ lbm}$$

The answer is B.

118. Type II cement develops its strength quickly.

The answer is C.

119. The term "saturated surface dry" is abbreviated "SSD." This refers to a state where the aggregates have been soaked and then allowed to drain of free water.

The answer is C.

120. Class F flyashes are lower in CaO (lime) and higher in C (carbon) than class C flyashes.

The answer is B.

Solutions
Transportation

121. The curve radius is

$$R = \frac{(360°)(100 \text{ ft})}{2\pi D} = \frac{(360°)(100 \text{ ft})}{2\pi(2°)}$$
$$= 2864.789 \text{ ft}$$

The interior angle is

$$I = 54°56'24'' - 32°15'18'' = 22°41'6'' = 22.685°$$

The tangent length is

$$T = R \tan\left(\frac{I}{2}\right) = (2864.789 \text{ ft}) \tan\left(\frac{22.685°}{2}\right)$$
$$= 574.652 \text{ ft}$$

The northing (i.e., change in the north dimension) is

$$\Delta N = T \cos(\text{bearing})$$
$$= (574.652 \text{ ft}) \cos 54°56'24''$$
$$= (574.652 \text{ ft}) \cos 54.94°$$
$$= 330.100 \text{ ft}$$

The easting (i.e., change in the east dimension) is

$$\Delta E = T \sin(\text{bearing})$$
$$= (574.652 \text{ ft}) \sin 54°56'24''$$
$$= (574.652 \text{ ft}) \sin 54.94°$$
$$= 470.382$$

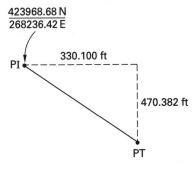

The PT is located at

$$(423968.68 + 330.100) \text{ N+}, \; (268236.42 + 470.382) \text{ E}$$
$$= 424298.78 \text{ N}, \; 268706.80 \text{ E}$$

The answer is A.

122. The bearing of the radius at the PT is offset 90° from the tangent at the PT.

$$180° - (90° + 54°56'24'') = 35°3'36''$$

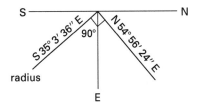

The answer is D.

123. The northing is

$$\Delta N = -R \cos(\text{bearing})$$
$$= (-2864.789 \text{ ft}) \cos 35°3'36''$$
$$= (-2864.789 \text{ ft}) \cos 35.06°$$
$$= -2344.976 \text{ ft}$$

The easting is

$$\Delta E = R \sin(\text{bearing})$$
$$= (2864.789 \text{ ft}) \sin 35°3'36''$$
$$= (2864.789 \text{ ft}) \sin 35.06°$$
$$= 1645.632 \text{ ft}$$

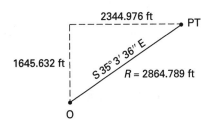

The center is located at

$$(424298.78 - 2344.976) \text{ N+}, \; (268706.80 + 1645.632) \text{ E}$$
$$= 421953.80 \text{ N}, \; 270352.43 \text{ E}$$

The answer is B.

124. The northing is

$$\Delta N = -T \cos (\text{bearing})$$
$$= -(574.652 \text{ ft}) \cos 32°15'18''$$
$$= -(574.652 \text{ ft}) \cos 32.255° = -485.972 \text{ ft}$$

The easting is

$$\Delta E = -T \sin (\text{bearing})$$
$$= -(574.652 \text{ ft}) \sin 32°15'18''$$
$$= -(574.652 \text{ ft}) \sin 32.255° = -306.685 \text{ ft}$$

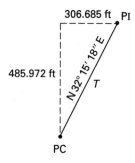

The PC is located at

$$(423968.68 - 485.972) \text{N}+, \ (268236.42 - 306.685) \text{E}$$
$$= 423482.71 \text{ N}, \ 267929.74 \text{ E}$$

The answer is C.

125. Use the Pythagorean theorem and the coordinates

$$d = \sqrt{\begin{array}{l}(424239.72 \text{ ft} - 424180.59 \text{ ft})^2 \\ +(268498.69 \text{ ft} - 268549.70 \text{ ft})^2\end{array}}$$
$$= 78.09 \text{ ft}$$

The answer is C.

126. The northing is

$$\Delta N = 424239.72 \text{ ft} - 421953.80 \text{ ft} = 2285.92 \text{ ft}$$

The easting is

$$\Delta E = 268498.69 \text{ ft} - 270352.43 \text{ ft} = -1853.74 \text{ ft}$$

The bearing is

$$\text{bearing} = \tan^{-1}\left(\frac{\Delta E}{\Delta N}\right)$$
$$= \tan^{-1}\left(\frac{-1853.74 \text{ ft}}{2285.92 \text{ ft}}\right)$$
$$= 39.04° = 39°2'24''$$

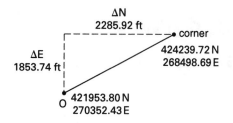

The difference in the bearings is

$$39°2'24'' - 35°3'26'' = 3°58'48''$$

The answer is D.

127. There are several rational ways to develop a reasonable estimate. One way is to base the answer on principles of mechanics.

The side friction factor is approximately

$$f_s = \frac{0.16 - 0.01(\text{v}_{\text{mph}} - 30)}{10}$$
$$= \frac{0.16 - 0.01(40 \text{ mph} - 30)}{10}$$
$$= 0.006$$

The theoretical superelevation is

$$e = \frac{\text{v}_{\text{mph}}^2}{15R} - f_s$$
$$= \frac{(40 \text{ mph})^2}{(15)(2864.789 \text{ ft})} - 0.006$$
$$= 0.031$$

If the 1990 AASHTO "Green Book" is used, assuming the maximum superelevation is about 0.08–0.10, then for 40 mph and a 2° curve, the superelevation is 0.027–0.028 (Tables III-10 and III- 11).

The answer is B.

128. There are several rational ways to develop a reasonable estimate. From the 1990 AASHTO "Green Book," Tables III-8 or III-9, the spiral length is 125 ft.

The answer is B.

129. Although even wider shoulders are desirable, particularly at higher speeds, a 10 ft clear zone is the minimum width needed to safely contain any stopping or off-course vehicles.

The answer is D.

130. From the 1990 AASHTO "Green Book" (Table III-1), the minimum stopping sight distance for a design speed of 40 mph is 275 to 325 ft.

The answer is B.

131. The degree of curve for roadway curves is

$$D = \frac{5729.6}{R} = \frac{5729.6}{2080 \text{ ft}}$$
$$= 2.755° \quad (2°45.3' \text{ or } 2°45'18'')$$

The answer is B.

132. The tangent length, T, is the distance from the PC to the PI.

$$T = R \tan\left(\frac{I}{2}\right) = (2080 \text{ ft}) \tan\left(\frac{60°}{2}\right)$$
$$= 1200.89 \text{ ft}$$

$$\text{sta PI} = \text{sta PC} + T = \text{sta } 12{+}40 + 1200.89 \text{ ft}$$
$$= \text{sta } 24{+}40.89$$

The answer is A.

133. The curve length is the distance from the PC to the PT.

$$L = \frac{2\pi R I}{360°} = \frac{2\pi(2080 \text{ ft})(60°)}{360°}$$
$$= 2178.2 \text{ ft}$$

$$\text{sta PT} = \text{sta PC} + L = \text{sta } 12{+}40 + 2178.17 \text{ ft}$$
$$= \text{sta } 34{+}18.17$$

The answer is C.

134. The chord length is

$$C = 2R \sin\left(\frac{I}{2}\right) = (2)(2080 \text{ ft}) \sin\left(\frac{60°}{2}\right)$$
$$= 2080 \text{ ft}$$

The answer is C.

135. The runout length, L_{runout}, as given in the table, is 150 ft for two lane roadways. All of this occurs in the curve. The tangent runout, T_R, is twice this (according to the problem statement) or 300 ft. The transition begins 300 ft before the beginning of the curve. Thus, $2/3$ of the transition occurs on the tangent and $1/3$ on the curve.

$$\text{sta transition start} = \text{sta PC} - T_R$$
$$= \text{sta } 12{+}40 - 300 \text{ ft}$$
$$= \text{sta } 9{+}40$$

The answer is A.

136. The radius of the curve is 2080 ft, which is between 2290 ft and 1910 ft in the table. Therefore, the superelevation rate is between 0.040 and 0.045.

The answer is C.

137. The total change in cross slope during the entire transition (runout and runoff) is $0.05 - (-0.02) = 0.07$. This occurs over a length of $L_{\text{runout}} + T_R = 150 \text{ ft} + 300 \text{ ft} = 450 \text{ ft}$. The change in cross slope from normal crown to reverse crown is $0.02 - (-0.02) = 0.04$. Normally, the rate of removal is the same as the superelevation runoff rate. Indeed, in this case, the transition rate was said to be constant. Therefore, the length from normal crown to reverse crown can be calculated from the ratio of total change in cross slope to total transition length.

$$\frac{0.07}{450 \text{ ft}} = \frac{0.04}{x}$$
$$x = \frac{(0.04)(450 \text{ ft})}{0.07}$$
$$= 257.14 \text{ ft}$$

$$\text{sta reverse crown} = \text{sta transition start} + x$$
$$= \text{sta } 9{+}40 + 257.14 \text{ ft}$$
$$= \text{sta } 11{+}97.14$$

The answer is B.

138. The midpoint, M, of the curve is $L/2$ past the PC.

$$\text{sta M} = \text{sta PC} + \frac{L}{2}$$
$$= \text{sta } 12{+}40 + \frac{2178.17 \text{ ft}}{2}$$
$$= \text{sta } 23{+}29.09$$

The elevation of the centerline at M is

$$\text{elev}_{M,\text{centerline}} = \text{elev}_{PC} + G\left(\frac{L}{2}\right)$$
$$= 170 \text{ ft} + \left(0.0075 \, \frac{\text{ft}}{\text{ft}}\right)\left(\frac{2178.17 \text{ ft}}{2}\right)$$
$$= 178.168 \text{ ft}$$

Since the midpoint is more than 150 ft from the PC (and more than 150 ft from the PT), the section is fully elevated. The cross slope is 0.048 ft/ft. The outside pavement edge is higher than the centerline by

$$\Delta_{\text{elev}} = \left(0.048 \, \frac{\text{ft}}{\text{ft}}\right)\left(12 \, \frac{\text{ft}}{\text{lane}}\right)(1 \text{ lane}) = 0.576 \text{ ft}$$

The elevation of the outside pavement edge is

$$\text{elev}_{M,\text{edge}} = \text{elev}_{M,\text{centerline}} + \Delta_{\text{elev}}$$
$$= 178.168 \text{ ft} + 0.576 \text{ ft}$$
$$= 178.744 \text{ ft}$$

The answer is D.

139.

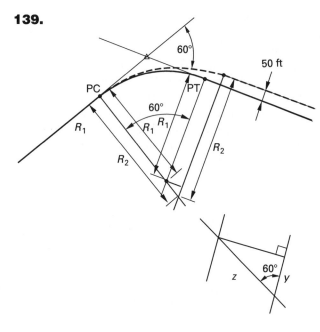

$$R_2 = R_1 + z = R_1 + y + 50 \text{ ft}$$

Subtracting R_1 from both sides,

$$z = y + 50 \text{ ft}$$

Also, $y/z = \cos 60° = \sin 30° = 1/2$, so, $y = z/2$.

$$z = \frac{z}{2} + 50 \text{ ft} = 100 \text{ ft}$$

$$y = 50 \text{ ft}$$

$$R_2 = R_1 + z = 2080 \text{ ft} + 100 \text{ ft}$$
$$= 2180 \text{ ft}$$

$$L_2 = \frac{2\pi R_2 I}{360°} = \frac{2\pi (2180 \text{ ft})(60°)}{360°}$$
$$= 2282.89 \text{ ft}$$

$$\text{sta PT}_2 = \text{sta PC} + L_2 = \text{sta } 12{+}40 + 2282.89 \text{ ft}$$
$$= \text{sta } 35{+}22.89$$

The answer is D.

140. $\quad T_2 = R_2 \tan\left(\frac{I}{2}\right) = (2180 \text{ ft})\tan\left(\frac{60°}{2}\right)$
$$= 1258.62 \text{ ft}$$

$$\text{sta PI}_2 = \text{sta PC} + T_2 = \text{sta } 12{+}40 + 1258.62 \text{ ft}$$
$$= \text{sta } 24{+}98.62$$

The answer is D.

141. There are three lanes of traffic in each direction on a six-lane freeway.

The answer is C.

142. The peak flow rate is the rate during the peak 15 min period expressed in vehicles per hour.

$$\begin{aligned}
\text{PHF} &= \frac{\text{actual hourly volume}_{\text{vph}}}{\text{peak rate of flow}_{\text{vph}}} \\
&= \frac{\text{actual hourly volume}_{\text{vph}}}{(4)(\text{peak 15 min volume})} \\
&= \frac{2400 \text{ vph}}{(4)(750 \text{ vph})} \\
&= 0.8
\end{aligned}$$

The answer is C.

143. The heavy vehicle factor is found in two steps. First, the passenger car equivalent, E, is found for each category of vehicle. Trucks and buses are combined into one group representing $14\% + 8\% = 22\%$ of the traffic.

From HCM-97, Table 3-3,

$$E_T = 1.5 \text{ (22\% trucks and buses)}$$

Since 22% is more than five times the percentage of recreational vehicles, all of the nonpassenger car traffic can be assumed to be trucks.

Calculate the heavy vehicle correction factor.

$$\begin{aligned}
f_{\text{HV}} &= \frac{1}{1 + P_T(E_T - 1)} \\
&= \frac{1}{1 + (0.22 + 0.04)(1.5 - 1)} \\
&= 0.885 \text{ veh/passenger car}
\end{aligned}$$

The answer is D.

144. Since the drivers are familiar with the route, the driver population adjustment is 1.0. The 15 min passenger-car equivalent flow rate is

$$\begin{aligned}
v_p &= \frac{V}{(\text{PHF})N f_{\text{HV}} f_p} \\
&= \frac{2400 \dfrac{\text{veh}}{\text{hr}}}{(0.8)(3 \text{ lanes})\left(0.885 \dfrac{\text{veh}}{\text{passenger car}}\right)(1.0)} \\
&= 1130 \text{ pcphpl}
\end{aligned}$$

The answer is B.

145. The maximum capacity of a highway at 70 mph is 2400 pcphpl. The volume/capacity ratio is

$$\frac{v}{c} = \frac{1130 \text{ pcphpl}}{2400 \text{ pcphpl}} = 0.471$$

The answer is A.

146. FFS

f_{LV}

f_{L}

f

f

FI f_{ID}

ph

The answ

147. Th HCM 97
Table 3-1

The answ

148. U the
LOS fro de-
crease t

The ans

149. I vice
flow ra

With

anged]

For a

v

$\frac{\text{veh}}{\text{senger car}}$ (1.0)

An vide the capacity.

Th

150. Use HCM-97 Table 3-1. When the LOS reaches D, the flow rate will be at 1440 (maximum for LOS C).

$$\text{future volume} = (\text{current volume})(1 + i)^t$$
$$1440 \text{ pcphpl} = (1130 \text{ pcphpl})(1 + 0.05)^t$$
$$1.274 = (1.05)^t$$
$$t = 4.97 \text{ years} \quad [\text{say 5 years}]$$

The answer is D.

151. The general equation of a parabolic curve is $y = x^2$, or $x = \sqrt{y}$. Therefore, the horizontal distance is proportional to the square root of the vertical distance, even if x and y have arbitrary reference points.

The answer is A.

152. The minimum elevation is

$$y_{\text{curve}} = y_{\text{rails}} + \text{clearance}$$
$$= 195.00 \text{ ft} + 26.00 \text{ ft}$$
$$= 221.00 \text{ ft}$$

The answer is D.

153. Refer to the following illustration.

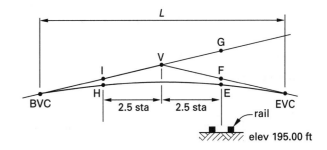

Horizontal distance x_{VG} = sta 28+50 − sta 26+00 = 2.5 sta.

The vertical offset distance EG is

$$y_{\text{EG}} = y_{\text{V}} + G_1 x_{\text{VG}} - y_{\text{E}}$$
$$= 231.00 \text{ ft} + \left(3 \; \frac{\text{ft}}{\text{sta}}\right)(2.5 \text{ sta}) - 221.00 \text{ ft}$$
$$= 17.50 \text{ ft}$$

Similarly, the vertical offset distance EF is

$$y_{\text{EF}} = y_{\text{V}} + G_2 x_{\text{VE}} - y_{\text{E}}$$
$$= 231.00 \text{ ft} + \left(-2 \; \frac{\text{ft}}{\text{sta}}\right)(2.5 \text{ sta}) - 221.00 \text{ ft}$$
$$= 5.00 \text{ ft}$$

Since this is an equal-tangent curve, and since points H and E are located equal horizontal distances from the vertex, distance HI and EF are equal.

$$y_{HI} = y_{EF} = 5.00 \text{ ft}$$

Also, the horizontal distance from the vertex to both the BVC and EVC are equal to half of the curve length, L.

$$x_V - x_{BVC} = x_{EVC} - x_V = \frac{L}{2}$$

Vertical offset distances from points on the curve to a tangent are proportional to the square of the horizontal distances from the BVC. Rather than find the constant of proportionality, which could also be done to solve this problem, use an equality of two vertical distances, HI and EF.

$$\frac{y_{EG}}{(x_E - x_{BVC})^2} = \frac{y_{HI}}{(x_H - x_{BVC})^2}$$

$$\frac{17.5 \text{ ft}}{\left(\dfrac{L}{2} + 2.5\right)^2} = \frac{5 \text{ ft}}{\left(\dfrac{L}{2} - 2.5\right)^2}$$

This reduces to a quadratic equation.

$$3.125L^2 - 56.25L + 78.125 = 0$$
$$L^2 - 18L + 25 = 0$$

The length of the curve is $L = 16.48$ sta.

The answer is C.

154. Once the curve length is known, the elevation of the BVC can be calculated.

$$y_{BVC} = y_V - G_1(x_V - x_{BVC}) = y_V - G_1\left(\frac{L}{2}\right)$$

$$= 231.00 \text{ ft} - \left(3 \frac{\text{ft}}{\text{sta}}\right)\left(\frac{16.48 \text{ sta}}{2}\right)$$

$$= 206.28 \text{ ft}$$

The answer is C.

155. The station of the BVC is

$$x_{BVC} = x_V - \frac{L}{2} = \text{sta } 26.00 - \frac{16.48 \text{ sta}}{2}$$

$$= \text{sta } 17.76$$

The answer is B.

156. The station of the EVC is

$$x_{EVC} = x_V + \frac{L}{2} = \text{sta } 26.00 + \frac{16.48 \text{ sta}}{2}$$

$$= \text{sta } 34.24$$

The answer is D.

157. The elevation of the EVC is

$$y_{EVC} = y_V + G_2(x_{EVC} - x_V) = y_V + G_2\left(\frac{L}{2}\right)$$

$$= 231.00 \text{ ft} + \left(-2 \frac{\text{ft}}{\text{sta}}\right)\left(\frac{16.48 \text{ sta}}{2}\right)$$

$$= 214.52 \text{ ft}$$

The answer is B.

158. The rate of grade change per station is

$$R = \frac{G_2 - G_1}{L} = \frac{-2 \dfrac{\text{ft}}{\text{sta}} - 3 \dfrac{\text{ft}}{\text{sta}}}{16.48 \text{ sta}}$$

$$= -0.3034 \text{ ft/sta}^2$$

The distance from the BVC to the turning point is

$$x_{TP} - x_{BVC} = \frac{-G_1}{R} = \frac{-3 \dfrac{\text{ft}}{\text{sta}}}{-0.3034 \dfrac{\text{ft}}{\text{sta}^2}}$$

$$= \text{sta } 9.89$$

The station of the turning point is

$$x_{TP} = (x_{TP} - x_{BVC}) + x_{BVC}$$
$$= \text{sta } 17.76 + \text{sta } 9.89$$
$$= \text{sta } 27.65$$

The answer is D.

159. Assembling the equation from previous results and measuring x in stations, the equation is

$$y = \left(\frac{R}{2}\right) x^2 + G_1 x + y_{BVC}$$

$$= \left(\frac{-0.3034}{2}\right) x^2 + 3x + 206.28 \text{ ft}$$

$$= -0.1517x^2 + 3x + 206.28 \text{ ft}$$

The distance from the BVC to the turning point is $x = 9.89$ sta.

$$y = \left(\frac{-0.3034 \ \frac{\text{ft}}{\text{sta}^2}}{2} \right) (9.89 \text{ sta})^2$$

$$+ \left(3 \ \frac{\text{ft}}{\text{sta}} \right) (9.89 \text{ sta}) + 206.28 \text{ ft}$$

$$= 221.11 \text{ ft}$$

The answer is A.

160. As determined in the previous question,

$$y = \left(\frac{R}{2} \right) x^2 + G_1 x + y_{\text{BVC}}$$

$$= (-0.1517)x^2 + 3x + 206.28 \text{ ft}$$

The answer is C.

Solutions
Environmental

161. The BOD loading is

$$L_{\text{BOD}} = \frac{S_0 Q}{A}$$

$$= \frac{\left(235\frac{\text{mg}}{\text{L}}\right)\left(0.32\frac{\text{MG}}{\text{day}}\right)\left(8.345\frac{\text{lbm-L}}{\text{mg-MG}}\right)}{30 \text{ ac}}$$

$$= 20.92 \text{ lbm/ac-day}$$

The answer is B.

162. An average for BOD loading is 0.20 lbm/person-day. (Other values are acceptable.) The population served is

$$P_e = \frac{L_{\text{BOD}}}{p_e} = \frac{35\dfrac{\text{lbm}}{\text{ac-day}}}{0.20\dfrac{\text{lbm}}{\text{person-day}}}$$

$$= 175 \text{ people/ac}$$

The answer is B.

163. The increase in volume as the depth increases from 2 ft to 5 ft is

$$\Delta V = A\Delta d$$

$$= (30 \text{ ac})\left(43{,}560\frac{\text{ft}^2}{\text{ac}}\right)(5 \text{ ft} - 2 \text{ ft})$$

$$= 3{,}920{,}400 \text{ ft}^3$$

During the winter, the flow is

$$Q = \frac{\left(0.32\frac{\text{MG}}{\text{day}}\right)\left(1{,}000{,}000\frac{\text{gal}}{\text{MG}}\right)}{7.48\frac{\text{gal}}{\text{ft}^3}}$$

$$= 42{,}781 \text{ ft}^3/\text{day}$$

During the winter, the loss due to evaporation is

$$E = \frac{\left(0.08\frac{\text{in}}{\text{day}}\right)(30 \text{ ac})\left(43{,}560\frac{\text{ft}^2}{\text{ac}}\right)}{12\frac{\text{in}}{\text{ft}}}$$

$$= 8712 \text{ ft}^3/\text{day}$$

The time required to increase the water depth from 2 ft to 5 ft is

$$t = \frac{\Delta V}{Q - E}$$

$$= \frac{3{,}920{,}400 \text{ ft}^3}{42{,}781\dfrac{\text{ft}^3}{\text{day}} - 8712\dfrac{\text{ft}^3}{\text{day}}}$$

$$= 115 \text{ days}$$

The answer is D.

164. Although the details are sketchy, the following facts probably apply.

I: The plating waste will contain heavy metals. Not only is it possible that the ponds will not be able to process the plating waste, but the metallic compounds might also inhibit biological growth in the ponds.

II: Plating waste is an ecological danger. Containment ponds must be constructed to rigorous standards that might not have been used to originally construct the town's pond.

III: Metallic compounds may be colored, but most are transparent. The fact that the facility proposes for the plating waste to bypass any sedimentation processing would indicate that suspended solids content is not an issue.

IV: The toxic effect of the plating waste would probably decrease pond vegetation and vectors.

V: It is unlikely that plating waste would find its way into the drinking water mains. The dedicated line is monitored. If plating waste from pond seepage could enter the water mains, then biological contamination from the current ponds would also be a problem.

The answer is A.

165. Although the details are sketchy, the following facts probably apply.

I: The meat packing waste will contain meat, blood, bone particles, and fat. Ponds are able to process such biological waste.

II: Meat packing waste is generally not an ecological danger. Assuming that capacity is not an issue, the construction of the ponds is probably adequate.

III: The organic material will contribute to increased cloudiness. In fact, the fat will rise to the surface and may even form a layer of scum. With a decrease in light, the efficiency will decrease.

IV: The meat packing waste would probably increase pond organisms.

V: It is unlikely that plating waste would find its way into the drinking water mains. The dedicated line is monitored. If meat packing waste from pond seepage could enter the water mains, then biological contamination from the current ponds would also be a problem.

The answer is C.

166. Since the primary clarifying system removes 100 mg/L of BOD_5, the concentration entering the primary clarifiers is

$$175 \frac{mg}{L} + 100 \frac{mg}{L} = 275 \frac{mg}{L}$$

The per-capita population equivalent is

$$
\begin{aligned}
p_e &= \frac{S_o Q}{P_e} \\
&= \frac{\left(275 \frac{mg}{L}\right)\left(1 \frac{MG}{day}\right)\left(8.345 \frac{lbm\text{-}L}{MG\text{-}mg}\right)}{13{,}500 \text{ people}} \\
&= 0.17 \text{ lbm/capita-day}
\end{aligned}
$$

The answer is C.

167. By its very name, biochemical oxygen demand represents the usage of oxygen. Since the flow enters the secondary process with no dissolved oxygen, the oxygen needed to satisfy the demand and the oxygen needed for the effluent concentration come from the aerator.

The ideal amount of oxygen supplied is

$$
\begin{aligned}
[O_2] &= \left(175 \frac{mg}{L} - 65 \frac{mg}{L}\right) + 3.6 \frac{mg}{L} \\
&= 113.6 \text{ mg/L} \\
\dot{m}_{O_2} &= [O_2]Q \\
&= \left(113.6 \frac{mg}{L}\right)\left(1.0 \frac{MG}{day}\right)\left(8.345 \frac{lbm\text{-}L}{MG\text{-}mg}\right) \\
&= 948.0 \text{ lbm/day}
\end{aligned}
$$

The answer is B.

168. Air is approximately 23.2% oxygen by weight. The density of air is approximately 0.075 lbm/ft^3 at normal 1 atm and 70°F. The ideal volume of air is

$$
\begin{aligned}
V_{ideal} &= \frac{\dot{m}}{\rho} \\
&= \frac{1000 \frac{lbm}{day}}{(0.232)\left(0.075 \frac{lbm}{ft^3}\right)\left(24 \frac{hr}{day}\right)\left(60 \frac{min}{hr}\right)} \\
&= 39.9 \text{ ft}^3/\text{min}
\end{aligned}
$$

The answer is D.

169. The actual volume of air provided can be calculated from the transfer coefficient.

$$
\begin{aligned}
V_{actual} &= k_a Q [O_2] \\
&= \frac{\left(0.005 \frac{ft^3\text{-}L}{gal\text{-}mg}\right)\left(1 \frac{MG}{day}\right)}{\left(24 \frac{hr}{day}\right)\left(60 \frac{min}{hr}\right)} \\
&\quad \times \left(1{,}000{,}000 \frac{gal}{MG}\right)\left(113.6 \frac{mg}{L}\right) \\
&= 394.4 \text{ ft}^3/\text{min}
\end{aligned}
$$

The answer is A.

170. The oxygen transfer efficiency is

$$
\eta = \frac{V_{actual}}{V_{ideal}} = \frac{39.9 \frac{ft^3}{min}}{394.4 \frac{ft^3}{min}}
$$
$$
= 0.101 \quad (10\%)
$$

The answer is C.

171. A conventional lagoon is typically 4 ft deep or less.
$$88.25 \text{ ft} + 4 \text{ ft} = 92.25 \text{ ft}$$

The answer is A.

172. Aerated lagoons are typically 10 to 15 ft deep. [TSS-1997, 93.416(b)] A minimum freeboard distance of 2 ft is required. [TSS-1997, 93.415] With a depth of 10 ft, the surface elevation would be

$$88.25 \text{ ft} + 10 \text{ ft} - 2 \text{ ft} = 96.25 \text{ ft}$$

The lagoon could be deeper, but the answer choices provided would not allow for any freeboard.

The answer is C.

173. *Ten States' Standards* minimum slope is 3:1. [TSS-1997, Sec. 93.413]

The answer is C.

174. The length of the bottom is twice the width. The area of the bottom of the lagoon is

$$A_{\text{bottom}} = w_{\text{bottom}} L_{\text{bottom}} = 2w_{\text{bottom}}^2$$

$$= (0.75 \text{ ac}) \left(43{,}560 \ \frac{\text{ft}^2}{\text{ac}} \right)$$

$$w_{\text{bottom}} = 127.8 \text{ ft} \quad [\text{use } 128 \text{ ft}]$$

$$L_{\text{bottom}} = 2w = (2)(128 \text{ ft})$$
$$= 256 \text{ ft}$$

If the surface elevation is 94.25 ft, depth is

$$d = 94.25 \text{ ft} - 88.25 \text{ ft} = 6 \text{ ft}$$

Since the dike slope is 3:1, the surface area expands by $(3)(6 \text{ ft}) = 18 \text{ ft}$ on each side. So, the surface dimensions are

$$w_{\text{surface}} = 128 \text{ ft} + (2)(18 \text{ ft}) = 164 \text{ ft}$$
$$L_{\text{surface}} = 256 \text{ ft} + (2)(18 \text{ ft}) = 292 \text{ ft}$$

The surface area is

$$A_{\text{surface}} = \frac{(164 \text{ ft})(292 \text{ ft})}{43{,}560 \ \frac{\text{ft}^2}{\text{ac}}} = 1.10 \text{ ac}$$

The answer is D.

175. Since the dike slopes are straight, the average "plan" area will occur at a depth of 2 ft. Since the dike slope is 3:1, the area expands by $(3)(2 \text{ ft}) = 6 \text{ ft}$ on each side. So, the average plan dimensions are

$$w_{\text{average}} = 128 \text{ ft} + (2)(6 \text{ ft}) = 140 \text{ ft}$$
$$L_{\text{average}} = 256 \text{ ft} + (2)(6 \text{ ft}) = 268 \text{ ft}$$

The average plan area is

$$A_{\text{average}} = (140 \text{ ft})(268 \text{ ft}) = 37{,}520 \text{ ft}^2$$

The volume is

$$V = A_{\text{average}} d = (37{,}520 \text{ ft}^2)(4 \text{ ft}) \left(7.4805 \ \frac{\text{gal}}{\text{ft}^3} \right)$$
$$= 1{,}123{,}000 \text{ gal}$$

The answer is B.

176.
$$Q = \frac{V}{t}$$

$$= \frac{1{,}123{,}000 \text{ gal}}{(60 \text{ days}) \left(1{,}000{,}000 \ \frac{\text{gal}}{\text{MG}} \right)}$$

$$= 0.0187 \text{ MG/day}$$

The answer is C.

177. "SLR" is the surface loading rate. Assume a population equivalent of 0.17 lbm BOD_5 per person.

$$P_e = \frac{A \ (\text{SLR})}{p_e}$$

$$= \frac{(42{,}400 \text{ ft}^2) \left(40 \ \frac{\text{lbm}}{\text{ac}} \right)}{\left(43{,}560 \ \frac{\text{ft}^2}{\text{ac}} \right) \left(0.17 \ \frac{\text{lbm}}{\text{person}} \right)}$$

$$= 229 \text{ people}$$

The answer is C.

178. The BOD_5 in the effluent is

$$S_o = (1 - \eta) S_i$$
$$= (1.00 - 0.89) \left(195 \ \frac{\text{lbm}}{\text{day}} \right)$$
$$= 21.45 \text{ lbm/day}$$

The concentration is

$$C = \frac{S_o}{Q} = \frac{\left(21.45 \ \frac{\text{lbm}}{\text{day}} \right) \left(1{,}000{,}000 \ \frac{\text{gal}}{\text{MG}} \right)}{\left(90{,}000 \ \frac{\text{gal}}{\text{day}} \right) \left(8.345 \ \frac{\text{lbm-L}}{\text{mg-MG}} \right)}$$

$$= 28.56 \text{ mg/L}$$

The answer is A.

179. The concentration of total solids is

$$C_{\text{TS}} = \frac{m_{\text{sample}} - m_{\text{tare}}}{V_{\text{sample}}}$$

$$= \frac{(0.1133 \text{ g} - 0.0985 \text{ g}) \left(1000 \ \frac{\text{mg}}{\text{g}} \right) \left(1000 \ \frac{\text{mL}}{\text{L}} \right)}{50 \text{ mL}}$$

$$= 296 \text{ mg/L}$$

The answer is D.

180. The concentration of volatile solids is

$$C_{\text{VS}} = \frac{(0.1133 \text{ g} - 0.1074 \text{ g})\left(1000 \frac{\text{mg}}{\text{g}}\right)\left(1000 \frac{\text{mL}}{\text{L}}\right)}{50 \text{ mL}}$$
$$= 118 \text{ mg/L}$$

The fraction of volatile solids is

$$f_{\text{VS}} = \frac{C_{\text{VS}}}{C_{\text{TS}}} = \frac{118 \frac{\text{mg}}{\text{L}}}{296 \frac{\text{mg}}{\text{L}}}$$
$$= 0.399$$

The answer is C.

181. The waste is circulated in an anaerobic process, but no air is added. Anaerobic digestion takes place in the absence of oxygen.

The answer is B.

182. The retention time for high-rate anaerobic digesters varies from 15 days to 25 days.

The answer is B.

183. The digester volume must be able to handle the waste flow for the retention period selected. This should be increased approximately 15% for freeboard volume.

$$V_{\text{total}} = Qt$$
$$= (1.15)\left(0.2 \frac{\text{ft}^3}{\text{sec}}\right)(15 \text{ days})$$
$$\times \left(24 \frac{\text{hr}}{\text{day}}\right)\left(3600 \frac{\text{sec}}{\text{hr}}\right)$$
$$= 298{,}080 \text{ ft}^3$$

Other methods of sizing the digestion volume are valid and can be used, particularly those based on BOD or COD. However, flow capacity determines the minimum value.

The answer is D.

184. As in most treatment operations, a minimum of two units is required. An even larger number will provide more flexibility in managing the digestion process. The actual number used will depend on standard designs and design specifications.

Traditional cylindrical anaerobic digesters have maximum depths of 20 to 45 ft. Diameters range from about 2 to 3 times the maximum depth. Assume digesters that are 25 ft deep (including freeboard) and 75 ft in diameter. (Other dimensions can be assumed. Smaller diameters will result in a larger number of required digesters.) The volume per unit is

$$V_{\text{digester}} = HA = H\left(\frac{\pi}{4}\right)D^2$$
$$= (25 \text{ ft})\left(\frac{\pi}{4}\right)(75 \text{ ft})^2$$
$$= 110{,}447 \text{ ft}^3$$

The number of tanks required is

$$n = \frac{V_{\text{total}}}{V_{\text{digester}}} = \frac{298{,}080 \text{ ft}^3}{110{,}447 \frac{\text{ft}^3}{\text{tank}}}$$
$$= 2.7 \text{ tanks}$$

At least three (maybe four) tanks should be built. The dimensions can be adjusted slightly to provide the total required volume.

The answer is D.

185. The flow rate is

$$Q = \left(0.2 \frac{\text{ft}^3}{\text{sec}}\right)\left(0.3048 \frac{\text{m}}{\text{ft}}\right)^3\left(24 \frac{\text{hr}}{\text{day}}\right)\left(3600 \frac{\text{sec}}{\text{hr}}\right)$$
$$= 489.3 \text{ m}^3/\text{day}$$

The ultimate BOD is

$$S_o = \frac{\left(8900 \frac{\text{mg}}{\text{L}}\right)\left(1000 \frac{\text{L}}{\text{m}^3}\right)}{\left(1000 \frac{\text{mg}}{\text{g}}\right)\left(1000 \frac{\text{g}}{\text{kg}}\right)} = 8.9 \text{ kg/m}^3$$

The cell tissue production rate is calculated from the yield constant.

$$P_x = kS_oQ$$
$$= \left(0.06 \frac{\text{g}}{\text{g}}\right)\left(8.9 \frac{\text{kg}}{\text{m}^3}\right)\left(489.3 \frac{\text{m}^3}{\text{day}}\right)$$
$$= 261.3 \text{ kg/day}$$

The answer is C.

186. The volume of methane produced is

$$V_{\text{methane}} = 0.35(EQS_o - 1.42P_x)$$

$$= \left(0.35 \, \frac{\text{m}^3}{\text{kg}}\right)$$

$$\times \left(\begin{array}{c} (0.70)\left(489.3 \, \frac{\text{m}^3}{\text{day}}\right)\left(8.9 \, \frac{\text{kg}}{\text{m}^3}\right) \\ - (1.42)\left(261.3 \, \frac{\text{kg}}{\text{day}}\right) \end{array}\right)$$

$$= 937 \text{ m}^3/\text{day}$$

The constant 0.35 m^3/kg was derived from tests with sucrose. It should be especially appropriate for this sugary waste.

The answer is B.

187. The lower heating value of dry methane gas is approximately 900 Btu/ft^3.

The answer is A.

188. 95°F to 98°F is the optimum temperature range for anaerobic digestion, and most digesters work in the 85°F to 100°F range.

The answer is A.

189. The bulk properties of the waste are not known. However, the waste is primarily water, so the specific heat and density of water will be used. The waste must be heated from 65°F to approximately 95°F. The total heat required is

$$q = Q_{\text{waste}}\rho_{\text{waste}}c_p(T_2 - T_1)$$

$$= \left(0.2 \, \frac{\text{ft}^3}{\text{sec}}\right)\left(3600 \, \frac{\text{sec}}{\text{hr}}\right)\left(24 \, \frac{\text{hr}}{\text{day}}\right)\left(62.4 \, \frac{\text{lbm}}{\text{ft}^3}\right)$$

$$\times \left(1.0 \, \frac{\text{Btu}}{\text{lbm-°F}}\right)(95°\text{F} - 65°\text{F})$$

$$= 3.23 \times 10^7 \text{ Btu/day}$$

The methane lower heating value is approximately 900 Btu/ft^3. The required methane volume is

$$V = \frac{q}{\text{LHV}} = \frac{3.23 \times 10^7 \, \frac{\text{Btu}}{\text{day}}}{900 \, \frac{\text{Btu}}{\text{ft}^3}}$$

$$= 35{,}900 \text{ ft}^3/\text{day}$$

The answer is C.

190. Methane production is essentially constant in the normal pH range of 6.5 to 8. However, carbonic acid (from CO_2) and organic acids (by-products of the initial stages of anaerobic digestion) decrease the pH. A highly alkaline waste will be able to buffer greater amounts of acid.

The answer is A.

191. 20% of the delivered solid waste (5% moisture and 15% classified materials) is removed prior to entering the furnace. The balance is 180 tons/day per power module. There are three power modules. The solid waste being brought to the plant is

$$W_{\text{waste}} = \frac{(3 \text{ modules})\left(180 \, \frac{\text{tons}}{\text{day-module}}\right)}{1 - 0.20}$$

$$= 675 \text{ tons/day}$$

The answer is D.

192. Each of the three modules receives 190 tons of sludge per day. The facility operates with an 85% utilization factor over 365 days per year.

$$W_{\text{sludge}} = (0.85)(3 \text{ modules})\left(190 \, \frac{\text{tons}}{\text{day-module}}\right)$$

$$\times \left(365 \, \frac{\text{days}}{\text{yr}}\right)$$

$$= 1.77 \times 10^5 \text{ tons/yr}$$

The answer is C.

193. Assume that each person generates 7 lbm of solid waste per day. (Other assumed values between 5 and 8 lbm will also yield the correct answer.) The population equivalent is

$$P_e = \frac{m}{p_e}$$

$$= \frac{\left(675 \, \frac{\text{tons}}{\text{day}}\right)\left(2000 \, \frac{\text{pounds}}{\text{ton}}\right)}{7 \, \frac{\text{pounds}}{\text{person-day}}}$$

$$= 1.93 \times 10^5 \text{ people}$$

The answer is C.

194. The dry weight of sludge solids from a mixed primary settling and trickling filter process is approximately 0.16 lbf/person-day. The total weight is

$$W_{\text{sludge solids}} = \frac{\left(0.16 \, \dfrac{\text{pounds}}{\text{person-day}}\right)(1.93 \times 10^5 \text{ people})}{2000 \, \dfrac{\text{pounds}}{\text{ton}}}$$

$$= 15.44 \text{ tons/day}$$

The answer is B.

195. If the total solids weight is 15.44 tons per day out of a total sludge weight of 190 tons per day per module, the solids content is

$$s = \frac{15.44 \, \dfrac{\text{tons}}{\text{day}}}{(3 \text{ modules})\left(190 \, \dfrac{\text{tons}}{\text{day-module}}\right)} = 0.0271$$

The answer is B.

196. The weight of solid waste received each year is

$$(0.85)\left(675 \, \frac{\text{tons}}{\text{day}}\right)\left(365 \, \frac{\text{days}}{\text{yr}}\right) = 2.09 \times 10^5 \text{ tons/yr}$$

From economic analysis tables, the $(A/P, 8\%, 20)$ equal-series present worth factor is 0.1019. The annualized construction cost per ton of waste is

$$A_1 = \frac{(\$11,000,000)(0.1019)}{2.09 \times 10^5 \text{ tons}} = \$5.36/\text{ton}$$

The answer is D.

197. The cost elements are the construction, labor, maintenance, and insurance costs. The annual cost of the labor, maintenance, and insurance per ton of waste received is

$$A_2 = \frac{\$600,000 + \$700,000 + \$100,000}{2.09 \times 10^5 \text{ tons}} = \$6.70/\text{ton}$$

The total equivalent annual cost per ton of waste received is

$$E = A_1 + A_2 = \$5.36 \, \frac{1}{\text{ton}} + \$6.70 \, \frac{1}{\text{ton}}$$

$$= \$12.06/\text{ton}$$

The answer is A.

198. The electrical revenue is

$$R_1 = \frac{(3 \text{ modules})\left(4000 \, \dfrac{\text{kW}}{\text{module}}\right) \times \left(24 \, \dfrac{\text{hr}}{\text{day}}\right)\left(\$0.015 \, \dfrac{1}{\text{kW-hr}}\right)}{675 \, \dfrac{\text{tons}}{\text{day}}}$$

$$= \$6.40/\text{ton}$$

The answer is B.

199. The recycling revenue per ton on incoming waste is

$$R_2 = (0.15)\left(\$2 \, \frac{1}{\text{ton}}\right)$$
$$+ (0.05)\left(\$6 \, \frac{1}{\text{ton}}\right)$$
$$+ (0.05)\left(\$50 \, \frac{1}{\text{ton}}\right)$$
$$+ (0.003)\left(\$350 \, \frac{1}{\text{ton}}\right)$$
$$+ (0.003)\left(\$400 \, \frac{1}{\text{ton}}\right)$$
$$= \$5.35/\text{ton}$$

The revenue from incinerating the sludge solids is

$$R_3 = \frac{\left(15.44 \, \dfrac{\text{tons}}{\text{day}}\right)\left(\$50 \, \dfrac{1}{\text{ton}}\right)}{(3 \text{ modules})\left(190 \, \dfrac{\text{tons}}{\text{day-module}}\right)}$$

$$= \$1.35/\text{ton}$$

The total revenue is

$$R = R_1 + R_2 + R_3$$
$$= \$6.40 \, \frac{1}{\text{ton}} + \$5.35 \, \frac{1}{\text{ton}} + \$1.35 \, \frac{1}{\text{ton}}$$
$$= \$13.10/\text{ton}$$

The answer is C.

200. The net profit of operation per ton of waste received is

$$P = R - E = \frac{\$13.10}{\text{ton}} - \frac{\$12.06}{\text{ton}}$$
$$= \$1.04/\text{ton}$$

The answer is B.

Solutions
Geotechnical

201. For a sample with 50% consolidation, the time factor is

$$T_v = \tfrac{1}{4}\pi U_z^2 = \tfrac{1}{4}\pi(0.5)^2$$
$$= 0.196$$

Since the test was "double-drainage," the 2 in sample drained from its top and bottom surfaces while being compressed. The maximum distance that the water had to travel to escape from the sample was

$$H_d = \frac{t}{2} = \frac{2 \text{ in}}{2} = 1 \text{ in}$$

The coefficient of consolidation is

$$C_v = \frac{T_v H_d^2}{t}$$
$$= \frac{(0.196)(1 \text{ in})^2 \left(60 \dfrac{\text{min}}{\text{hr}}\right)}{(20 \text{ min}) \left(144 \dfrac{\text{in}^2}{\text{ft}^2}\right)}$$
$$= 0.00408 \text{ ft}^2/\text{hr}$$

The answer is B.

202. The traditional slope of layer thickness (or void ratio) is negative, since the layer is being compressed. Sand will drain and compress more quickly than clay, so its slope is steeper (i.e., more negative).

The answer is D.

203. The traditional slope of layer thickness (or void ratio) is negative, since the layer is being compressed. However, the rate of consolidation decreases, so the second derivative is positive. Sand essentially does not have any secondary consolidation, so the graph is horizontal. The second derivative is zero.

The answer is A.

204. For a 90% degree of consolidation, the time factor is approximately $T_v = 0.848$. Since the water in the clay cannot drain through the impervious layer, this is single-drainage. The maximum distance the water must travel to drain is the full clay layer.

The time to achieve primary consolidation is

$$t = \frac{T_v H_d^2}{C_v}$$
$$= \frac{(0.848)(10 \text{ ft})^2}{\left(0.00408 \dfrac{\text{ft}^2}{\text{hr}}\right)\left(24 \dfrac{\text{hr}}{\text{day}}\right)\left(365 \dfrac{\text{days}}{\text{yr}}\right)}$$
$$= 2.37 \text{ yr}$$

The answer is C.

205. The illustration must be redrawn to scale. Then, flow and equipotential lines can be drawn. While large amounts of time can be taken to get the flow net almost perfect, there is no need to do so in this problem, as the answer choices are significantly different.

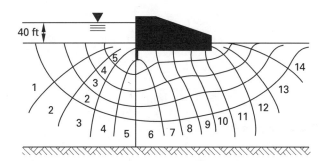

With five flow channels and fourteen equipotential drops, the flow rate is approximately

$$Q = KH \left(\frac{N_f}{N_p}\right)$$
$$= \left(0.005 \frac{\text{gal}}{\text{day-ft}^2}\right)(40 \text{ ft})\left(\frac{5}{14}\right)$$
$$= 0.071 \text{ gal/day-ft}$$

While the cutoff wall disturbs the flow net slightly, the effect past the second wall does not significantly "pinch" down the seepage. The effect of the wall is probably to reduce the flow less than 20%.

The answer is C.

206. The mass of wet sample 4 is

$$m_{4,\text{wet}} = (6330 \text{ g} - 4200 \text{ g}) \left(0.002205 \, \frac{\text{lbm}}{\text{g}} \right)$$

$$= 4.697 \text{ lbm}$$

The wet density is

$$\rho_{4,\text{wet}} = \frac{m_{4,\text{wet}}}{V} = \frac{4.697 \text{ lbm}}{\frac{1}{30} \text{ ft}^3}$$

$$= 140.9 \text{ lbm/ft}^3$$

The answer is A.

207. The mass of wet sample 1 is

$$m_{1,\text{wet}} = (6100 \text{ g} - 4200 \text{ g}) \left(0.002205 \, \frac{\text{lbm}}{\text{g}} \right)$$

$$= 4.190 \text{ lbm}$$

The wet density is

$$\rho_{1,\text{wet}} = \frac{m_{1,\text{wet}}}{V} = \frac{4.190 \text{ lbm}}{\frac{1}{30} \text{ ft}^3}$$

$$= 125.7 \text{ lbm/ft}^3$$

The dry density is

$$\rho_{1,\text{dry}} = \frac{\rho_{1,\text{wet}}}{1 + w_1} = \frac{125.7 \, \frac{\text{lbm}}{\text{ft}^3}}{1 + 0.082}$$

$$= 116.2 \text{ lbm/ft}^3$$

The answer is B.

208. The wet density of wet sample 2 is

$$\rho_{2,\text{wet}} = \frac{m_{2,\text{wet}}}{V}$$

$$= \frac{(6300 \text{ g} - 4200 \text{ g}) \left(0.002205 \, \frac{\text{lbm}}{\text{g}} \right)}{\frac{1}{30} \text{ ft}^3}$$

$$= 138.9 \text{ lbm/ft}^3$$

The dry density is

$$\rho_{2,\text{dry}} = \frac{\rho_{3,\text{wet}}}{1 + w} = \frac{138.9 \, \frac{\text{lbm}}{\text{ft}^3}}{1 + 0.101}$$

$$= 126.2 \text{ lbm/ft}^3$$

In order to calculate the relative compaction, the maximum dry density must be known. From the sample data given, it appears that the maximum density is close to sample 3. (Graphing the dry densities is also an option.)

The wet density of wet sample 3 is

$$\rho_{3,\text{wet}} = \frac{m_{3,\text{wet}}}{V}$$

$$= \frac{(6425 \text{ g} - 4200 \text{ g}) \left(0.002205 \, \frac{\text{lbm}}{\text{g}} \right)}{\frac{1}{30} \text{ ft}^3}$$

$$= 147.2 \text{ lbm/ft}^3$$

The dry density is

$$\rho_{3,\text{dry}} = \frac{\rho_{3,\text{wet}}}{1 + w} = \frac{147.2 \, \frac{\text{lbm}}{\text{ft}^3}}{1 + 0.117}$$

$$= 131.8 \text{ lbm/ft}^3$$

[Graphing may show the maximum to be slightly higher than this value.]

The relative compaction of sample 2 is

$$\frac{126.2 \, \frac{\text{lbm}}{\text{ft}^3}}{131.8 \, \frac{\text{lbm}}{\text{ft}^3}} = 0.958 \quad (96\%)$$

The answer is D.

209. The volumes of soils with high plasticity indices (PI) vary greatly with changes in the moisture content.

The answer is D.

210. A hydrometer analysis indicates the grain size of fine-grained soils.

The answer is C.

211. The relative compaction is calculated from the bulk density.

$$\frac{\gamma_b}{SG_{\text{max}} \gamma_{\text{water}}} = \frac{145.2 \, \frac{\text{lbf}}{\text{ft}^3}}{(2.5) \left(62.4 \, \frac{\text{lbf}}{\text{ft}^3} \right)} = 0.931$$

The answer is B.

212. Thin layers require fewer passes to achieve the required relative density. Thick layers hold heat longer, allowing more time to roll. In practice, lifts of about 5 in are maximum.

The answer is B.

213. The three factors that influence the compaction the most are the mixture temperature, the layer thickness, and the number of roller passes.

The answer is A.

214. Normal highway pavement mixtures contain 4% to 6% asphalt.

The answer is B.

215. The nuclear gauge cannot differentiate between the hydrogen atoms in water and the hydrogen atoms in the asphalt cement hydrocarbons. When using the percentage moisture scale (button) on a nuclear gauge over an asphalt pavement, the reading indicates the fraction of hydrocarbons present in the asphalt. Because the paving temperature is so high, the actual moisture content is usually less than $\frac{1}{2}\%$.

The answer is D.

216. The bulk "volume" is

$$m_b = m_{\text{SSD}} - m_{\text{submerged}}$$
$$= 1215 \text{ g} - 715 \text{ g}$$
$$= 500 \text{ g} \quad [500 \text{ cm}^3]$$

The answer is B.

217. The bulk specific gravity is

$$G_b = \frac{m_{\text{air}}}{m_b} = \frac{1210 \text{ g}}{500 \text{ g}}$$
$$= 2.42$$

The answer is C.

218. The bulk unit weight is

$$\gamma_b = G_b \gamma_{\text{water}} = (2.42)\left(62.4 \ \frac{\text{lbf}}{\text{ft}^3}\right)$$
$$= 151.0 \text{ lbf/ft}^3$$

The answer is B.

219. The relative compaction is

$$\frac{\gamma_b}{\gamma_{\text{max}}} = \frac{G_b}{G_{\text{max}}} = \frac{2.42}{2.56}$$
$$= 0.945 \quad (94.5\%)$$

The answer is A.

220. The percent air voids P_a (or VTM) is

$$\text{VTM} = 100\%\left(1 - \frac{G_b}{G_{\text{max}}}\right) = 100\%(1 - 0.945)$$
$$= 5.5\%$$

The answer is C.

221. The Rankine theory predicts the coefficient of active earth pressure as

$$k_a = \frac{1 - \sin\phi}{1 + \sin\phi} = \frac{1 - \sin 35°}{1 + \sin 35°}$$
$$= 0.271$$

The answer is C.

222. The horizontal soil pressure increases linearly to

$$p_{a,h} = k_a \gamma H = (0.271)\left(130 \ \frac{\text{lbf}}{\text{ft}^3}\right)(21 \text{ ft})$$
$$= 739.8 \text{ lbf/ft}^2$$

The total active force on the stem is

$$R_{a,h,\text{stem}} = \tfrac{1}{2} p_{a,h} H_{\text{stem}}$$
$$= \left(\frac{1}{2}\right)\left(739.8 \ \frac{\text{lbf}}{\text{ft}^2}\right)(21 \text{ ft})$$
$$= 7768 \text{ lbf/ft}$$

The answer is A.

223. The horizontal force on the stem acts one-third of the stem height above the top of the footing base. The moment arm (measured from the top of the footing base) is

$$y = \frac{H_{\text{stem}}}{3} = \frac{21 \text{ ft}}{3} = 7 \text{ ft}$$
$$M_{a,\text{stem}} = y R_{a,h,\text{stem}}$$
$$= (7 \text{ ft})\left(7768 \ \frac{\text{lbf}}{\text{ft}}\right)$$
$$= 54{,}376 \text{ ft-lbf/ft}$$

The answer is A.

224. Since there is no passive force, the shear on the stem is the same as the horizontal active force, 7768 lbf/ft.

The answer is B.

225. The overturning moment includes the moment of the active force on the end of the base. The active resultant acting on the full 23 ft height is

$$R_a = \tfrac{1}{2}k_a\gamma H^2$$

$$= \left(\frac{1}{2}\right)(0.271)\left(130\ \frac{\text{lbf}}{\text{ft}^3}\right)(23\ \text{ft})^2$$

$$= 9318\ \text{lbf/ft}$$

This resultant acts $H/3 = 23$ ft/3 up from the bottom of the base. Taking moments about the toe, the overturning moment is

$$M_{\text{overturning}} = yR_a = \left(\frac{23\ \text{ft}}{3}\right)\left(9318\ \frac{\text{lbf}}{\text{ft}}\right)$$

$$= 71{,}438\ \text{ft-lbf/ft}$$

The answer is D.

226. The stem, base, and backfill produce the stabilizing moment. Since the specific weight of concrete was not given in this problem, it is appropriate to use the "standard" value of 150 lbf/ft^3 for reinforced concrete. The weight of the base per foot of wall is

$$W_{\text{base}} = \gamma_{\text{concrete}}V_{\text{base}}$$

$$= \left(150\ \frac{\text{lbf}}{\text{ft}^3}\right)(2\ \text{ft})(12\ \text{ft})$$

$$= 3600\ \text{lbf/ft}$$

The horizontal distance from the toe to this force is

$$x_{\text{base}} = \frac{B}{2} = \frac{12\ \text{ft}}{2} = 6\ \text{ft}$$

The weight of the stem per foot of wall is

$$W_{\text{stem}} = \gamma_{\text{concrete}}V_{\text{stem}}$$

$$= \left(150\ \frac{\text{lbf}}{\text{ft}^3}\right)(1.75\ \text{ft})(21\ \text{ft})$$

$$= 5513\ \text{lbf/ft}$$

The horizontal distance from the toe to this force is

$$x_{\text{stem}} = 4.25\ \text{ft} + \frac{1.75\ \text{ft}}{2} = 5.125\ \text{ft}$$

The weight of the active soil above the base heel is

$$W_{\text{soil}} = \gamma_{\text{soil}}V_{\text{soil}}$$

$$= \left(130\ \frac{\text{lbf}}{\text{ft}^3}\right)(21\ \text{ft})(6\ \text{ft})$$

$$= 16{,}380\ \text{lbf/ft}$$

The horizontal distance from the toe to the centroid of the backfill soil is

$$x_{\text{soil}} = 12\ \text{ft} - \frac{6\ \text{ft}}{2} = 9\ \text{ft}$$

The stabilizing (i.e., resisting) moment is

$$M_{\text{resisting}} = \sum xW$$

$$= (6\ \text{ft})\left(3600\ \frac{\text{lbf}}{\text{ft}}\right)$$

$$+ (5.125\ \text{ft})\left(5513\ \frac{\text{lbf}}{\text{ft}}\right)$$

$$+ (9\ \text{ft})\left(16{,}380\ \frac{\text{lbf}}{\text{ft}}\right)$$

$$= 197{,}274\ \text{ft-lbf/ft}$$

The answer is D.

227. The factor of safety against overturning is

$$F_{\text{OT}} = \frac{M_{\text{resisting}}}{M_{\text{overturning}}} = \frac{197{,}274\ \frac{\text{ft-lbf}}{\text{ft}}}{71{,}438\ \frac{\text{ft-lbf}}{\text{ft}}}$$

$$= 2.76$$

The answer is C.

228. The total vertical weight on the base soil is

$$W_{\text{total}} = W_{\text{base}} + W_{\text{stem}} + W_{\text{soil}}$$

$$= 3600\ \frac{\text{lbf}}{\text{ft}} + 5513\ \frac{\text{lbf}}{\text{ft}} + 16{,}380\ \frac{\text{lbf}}{\text{ft}}$$

$$= 25{,}493\ \text{lbf/ft}$$

The friction force resisting sliding is

$$R_{\text{SL}} = fW_{\text{total}} = (0.60)\left(25{,}493\ \frac{\text{lbf}}{\text{ft}}\right)$$

$$= 15{,}296\ \text{lbf/ft}$$

The factor of safety against sliding is

$$F_{\text{SL}} = \frac{R_{\text{SL}}}{R_a} = \frac{15{,}296\ \frac{\text{lbf}}{\text{ft}}}{9318\ \frac{\text{lbf}}{\text{ft}}}$$

$$= 1.64$$

The margin of safety is 1 less than the factor of safety.

$$MS = F - 1 = 1.64 - 1 = 0.64$$

The answer is A.

229. The eccentricity is the horizontal distance between the midpoint of the base and the upward vertical resultant. The downward vertical resultant was calculated as the total vertical weight. Measuring distance from the toe, the location of the vertical resultant is

$$x = \frac{M_{\text{resisting}} - M_{\text{overturning}}}{W_{\text{total}}}$$

$$= \frac{197{,}274 \,\frac{\text{ft-lbf}}{\text{ft}} - 71{,}438 \,\frac{\text{ft-lbf}}{\text{ft}}}{25{,}493 \,\frac{\text{lbf}}{\text{ft}}}$$

$$= 4.936 \text{ ft}$$

The eccentricity is

$$\epsilon = \frac{B}{2} - x = \frac{12 \text{ ft}}{2} - 4.936 \text{ ft}$$

$$= 1.064 \text{ ft}$$

The pressure will be maximum at the toe.

$$p_{\max} = \left(\frac{W_{\text{total}}}{B}\right)\left(1 + \frac{6\epsilon}{B}\right)$$

$$= \left(\frac{25{,}493 \,\frac{\text{lbf}}{\text{ft}}}{12 \text{ ft}}\right)\left(1 + \frac{(6)(1.064 \text{ ft})}{12 \text{ ft}}\right)$$

$$= 3255 \text{ lbf/ft}^2$$

The answer is D.

230. The pressure will be minimum at the heel. Since the resultant is within the middle third of the base, the soil will be in compression everywhere. The minimum pressure is

$$p_{\min} = \left(\frac{W_{\text{total}}}{B}\right)\left(1 - \frac{6\epsilon}{B}\right)$$

$$= \left(\frac{25{,}493 \,\frac{\text{lbf}}{\text{ft}}}{12 \text{ ft}}\right)\left(1 - \frac{(6)(1.064 \text{ ft})}{12 \text{ ft}}\right)$$

$$= 994 \text{ lbf/ft}^2$$

The answer is C.

231. The weight of solids in 1 ft^3 of compacted fill is

$$W_s = \frac{W_t}{1 + w} = \frac{110 \text{ lbf}}{1 + 0.17} = 94.02 \text{ lbf}$$

The answer is C.

232. Since the unit weight of the compacted fill is 110 lbf/ft^3, the weight of the water in 1 ft^3 of fill is

$$W_w = W_t - W_s$$
$$= 110 \text{ lbf} - 94.02 \text{ lbf}$$
$$= 15.98 \text{ lbf}$$

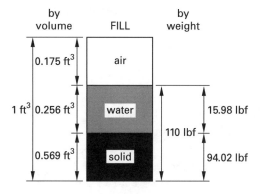

The answer is B.

233. The void ratio is

$$e = 0.70 = \frac{V_v}{V_s}$$

Therefore,

$$V_s = \frac{V_v}{0.70} = 1.429 V_v$$

The solid and void volumes make up the entire volume. For 1 ft^3 of borrow soil,

$$V_s + V_v = 1 \text{ ft}^3$$
$$1.429 V_v + V_v = 2.429 V_v = 1 \text{ ft}^3$$
$$V_v = 0.412 \text{ ft}^3$$

The answer is A.

234.
$$V_s = \frac{V_v}{0.70} = 1.429 V_v$$
$$= (1.429)(0.412 \text{ ft}^3)$$
$$= 0.589 \text{ ft}^3$$

The answer is D.

235. The degree of saturation is

$$s = 0.40 = \frac{V_w}{V_v}$$

$$V_w = 0.40V_v = (0.40)(0.412 \text{ ft}^3)$$

$$= 0.165 \text{ ft}^3$$

The answer is D.

236. The weight of solids per cubic foot can be found from the specific gravity.

$$W_s = V_s\gamma_s = V_s(\text{SG})\gamma_w$$

$$= (0.589 \text{ ft}^3)(2.65)\left(62.4 \frac{\text{lbf}}{\text{ft}^3}\right)$$

$$= 97.4 \text{ lbf}$$

The answer is C.

237. The weight of the water in 1 ft^3 of undisturbed borrow soil is

$$W_w = V_w\gamma_w = (0.165 \text{ ft}^3)\left(62.4 \frac{\text{lbf}}{\text{ft}^3}\right)$$

$$= 10.3 \text{ lbf}$$

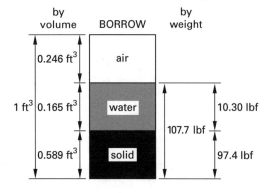

The answer is C.

238. The total weight of solids needed in the dam is

$$W_{s,\text{total}} = V\gamma$$

$$= (3 \times 10^6 \text{ yd}^3)\left(27 \frac{\text{ft}^3}{\text{yd}^3}\right)\left(94.02 \frac{\text{lbf}}{\text{ft}^3}\right)$$

$$= 7.616 \times 10^9 \text{ lbf}$$

Each cubic foot of borrow soil supplies 97.4 lbm of solids. The volume of borrow soil is

$$V_{\text{borrow}} = \frac{W_{s,\text{total}}}{W_s} = \frac{7.616 \times 10^9 \text{ lbf}}{\left(97.4 \frac{\text{lbf}}{\text{ft}^3}\right)\left(27 \frac{\text{ft}^3}{\text{yd}^3}\right)}$$

$$= 2.9 \times 10^6 \text{ yd}^3$$

The answer is B.

239. This question requires some judgment. When the borrow soil is dug out and dropped into the truck, it is not compacted. Expansive clay-type soil would not be a good choice for an earth-filled dam, so it can be assumed that the borrow soil does not come out of the pit in chunks (which might result in large voids in the transported soil). Furthermore, if 25% of the truck load was empty space, the truck would be greatly underutilized. 1% is too little, and 25% is too much. By elimination, the fluff factor is probably 15%.

The answer is C.

240. The unit weight of the borrow soil is

$$\gamma_w = W_s + W_w = 97.4 \frac{\text{lbf}}{\text{ft}^3} + 10.3 \frac{\text{lbf}}{\text{ft}^3}$$

$$= 107.7 \text{ lbf/ft}^3$$

Including a 10% fluff factor, the maximum weight that can be loaded onto a truck per trip is

$$W_{\text{truck}} = \frac{(10 \text{ yd}^3)\left(27 \frac{\text{ft}^3}{\text{yd}^3}\right)\left(107.7 \frac{\text{lbf}}{\text{ft}^3}\right)}{(1.10)\left(2000 \frac{\text{lbf}}{\text{ton}}\right)}$$

$$= 13.22 \text{ tons}$$

Since this is less than the maximum capacity of 14 tons, the number of truck loads is controlled by volume. The number of trips required is

$$n = \frac{V_{\text{borrow}}}{V_{\text{trip}}} = \frac{(1.10)(2.9 \times 10^6 \text{ yd}^3)}{10 \frac{\text{yd}^3}{\text{trip}}}$$

$$= 3.19 \times 10^5 \text{ trips}$$

The answer is D.

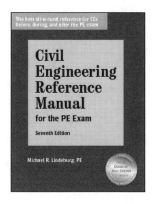
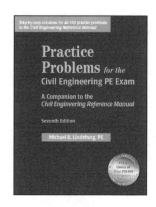
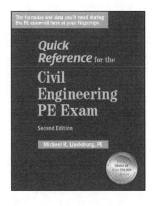
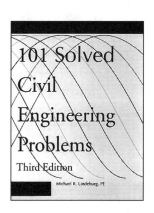

BUSINESS REPLY MAIL

FIRST CLASS MAIL   PERMIT NO. 33   BELMONT, CA

POSTAGE WILL BE PAID BY ADDRESSEE

PROFESSIONAL PUBLICATIONS INC
1250 FIFTH AVE
BELMONT CA 94002-9979

NO POSTAGE
NECESSARY
IF MAILED
IN THE
UNITED STATES

BUSINESS REPLY MAIL

FIRST CLASS MAIL   PERMIT NO. 33   BELMONT, CA

POSTAGE WILL BE PAID BY ADDRESSEE

PROFESSIONAL PUBLICATIONS INC
1250 FIFTH AVE
BELMONT CA 94002-9979

NO POSTAGE
NECESSARY
IF MAILED
IN THE
UNITED STATES

BUSINESS REPLY MAIL

FIRST CLASS MAIL   PERMIT NO. 33   BELMONT, CA

POSTAGE WILL BE PAID BY ADDRESSEE

PROFESSIONAL PUBLICATIONS INC
1250 FIFTH AVE
BELMONT CA 94002-9979